SUTTONS
Encyclopaedia of
VEGETABLES

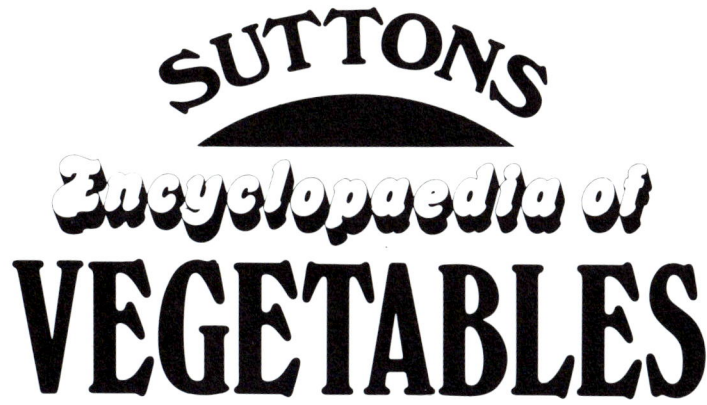

SUTTONS
Encyclopaedia of
VEGETABLES

FRED POTTER
V.M.H., A.H.R.H.S.

and

FRANK SHACKEL
A.H.R.H.S.

Pelham Books

First published in Great Britain by
PELHAM BOOKS LTD
52 Bedford Square
London WC1B 3EF
1976

ISBN 0 7207 0923 7

Filmset and Printed in Great Britain by
BAS Printers Limited, Wallop, Hampshire
and bound by Dorstel Press, Harlow, Essex

Contents

List of Illustrations

Foreword
by O. P. F. Sutton

I was delighted to be asked to write this Foreword, as I have long thought that a fully comprehensive book on vegetable-growing would be of great assistance to the vast numbers of amateur gardeners. During the year I receive a variety of enquiries in our business seeking advice on cultivations under glass or in the open, growing for table or freezing purposes and also for exhibition. The co-authors have both been specializing in vegetable seed production for over fifty years and I have known them personally throughout my business life—forty-four years or so. I am therefore in a position to appreciate their unrivalled knowledge of vegetable species and varieties, and also their unfailing efforts to impart the benefit of their practical knowledge not only to the amateur but also to the commercial grower.

For many years they have both served on the Fruit and Vegetable Committee of the Royal Horticultural Society, who have recognized their experience and ability by making distinguished awards for their services to horticulture. They are experienced judges at horticultural shows over a wide area and during the winter are much sought after to give lectures in many parts of the country.

Both authors have travelled extensively in Europe and have studied Continental varieties of vegetables and their uses under varying climatic and cultural conditions. They have frequently been successful in introducing to this country entirely new varieties which have ultimately become well known throughout the British Isles. As a result of their breeding efforts varieties such as the popular Exhibition Pea 'Show Perfection', Runner Bean 'Achievement' and many others have been produced.

This book differs in many respects from others on the same subject. For example, the grouping of the complex Cauliflower Section has never before to my knowledge been accomplished in such a clear manner. It not only contains the history of the various groups but it also assists the grower in finding the correct type for a specific purpose or time of cutting. Similarly, the grouping of Lettuces and the approach to setting out the cultivation methods of each section of this subject simplifies what can often be confusing.

Suttons Encyclopaedia of Vegetables deals with the growing of vegetables for fresh table use and freezing, growing under glass (heated or cold), frames, cloches, and polythene tunnels. Other sections include advice on the preparation and staging of vegetables for exhibition, insect pests and diseases (stressing prevention rather than cure), monthly reminders and, last but not least, the importance of understanding your soil, its preparation and its feeding to maintain and improve the physical condition for better crops.

I am sure this book will be of untold value to all growers of vegetables for many years to come.

January 1976

Preface

Vegetables play an extremely important part in assisting to feed the family. Fresh home-grown produce has its distinct health-giving properties, and, given the 'Know how', vegetable gardening can and should be an excellent form of mental and physical relaxation.

During the First and Second World Wars home-produced vegetables played an important part in feeding the nation, and, at the time of writing this book, the economic conditions of the country as a whole have induced many households with land at their disposal to increase vegetable production. It was this consideration which partly prompted the writing of this book.

In the following chapters we have tried to cover every type of vegetable which can be grown successfully and profitably in this country. In addition we have tried to record our experience from a lifetime's work entirely in vegetables and have included a little history of both the species and the varieties.

We have long since appreciated that amateur growers support some of the many horticultural shows that are held throughout the country; we have therefore included in this book growing, preparing and staging for shows. At the same time, the problem of insect pests and diseases, their prevention or cure, has been incorporated.

The basic principles for a successful vegetable garden are as follows:

Learn to understand your soil as its texture can vary from clay to heavy or light loam, sand or gravel. With the heavier types of soil make sure that drainage is adequate as a badly drained soil will produced sour conditions which must be eradicated. With some types of soil, whilst the top spit may appear to be excellent, the sub-soil can form a pan, which should be broken but never brought to the surface. This will not only assist drainage but create a better-developed root system.

For clay and the heavier types of soil late autumn or early winter digging is essential. This allows sufficient time for the frost to pulverize the surface, which means that at sowing season the soil will crumble down after a few days of drying winds into a fine tilth ideal for seed sowing.

Manuring. Your use of the compost-heap to provide the necessary humus is essential, with certain reservations according to the crops being grown. Where it is difficult to obtain farmyard or stable manure, a profitable substitute is the homemade compost-heap which should be built up over a period of the whole year, incorporating such material as straw, household waste in the form of potato peelings, the outside leaves of lettuce and cabbage, pods of the various legumes, waste paper and any old woollen articles, all garden debris, but avoid such material as hard woods, privet hedge-clippings, rose-prunings, etc. When a reasonable thickness has been obtained in the compost-heap, a dusting with a proprietory brand of accelerator or sulphate ammonia will assist in the decomposition.

The use of fertilizers, particularly the organic types, are invaluable in the preparation of the ground and they are often used during the course of growth.

Prior to the winter preparation of the soil, plan the garden to suit the family requirements, bearing in mind that it is more important to make ample provision for the winter months rather than have a surplus of vegetables in the summer. When possible, think in terms of quick-maturing varieties sown in succession, little and often; this will mean young succulent produce at all times, occupying the ground for a short period only, thus producing a heavier weight of produce per annum from a given area. Study carefully your local climatic conditions, selecting varieties which are suitable, and adapt your sowing and planting times accordingly for outdoor crops.

If you possess a greenhouse, heated or cold, give careful thought to sowing dates, whether you are growing entirely under glass or raising plants for transplanting into the open ground. If the latter, particularly with biennial subjects, make quite certain they are well hardened before final translating into the open; in this way you may avoid what so often happens, plants bolting to seed in the early stages. This is due to vernalization; a few warms days after transplanting, followed by one or two cold nights induces the plant to complete its life cycle by running to seed.

Sowing Depth. For all small vegetable seeds the average depth is about 1 in.; whereas with the larger-seeded types, such as Peas and Beans, 2–3 in., according to soil conditions, is the maximum. Crops grown *in situ* require singling and therefore thinning out is essential; this operation should be carried out as soon as the seedlings are large enough to handle to avoid checks in plant growth. This applies particularly to such crops as Beets, Lettuce, Onions, Parsnip, Spinach, Turnips etc. Do not sow Brassicas (Cabbages, Brussels Sprouts) thickly; this will enable the individual plants to make good root action, which is essential before transplanting. When transplanting, adhere to the spacings recommended in the table at the end of the book.

General Cultivations during Growth. The hoe is without doubt the best friend of any gardener and, if used regularly, will prevent indigenous weeds taking command and create a tilth which will in turn preserve moisture, so important during a dry weather spell, and will prevent the surface from being panned down by heavy rains.

During the course of growth, remember that prevention of insect pest and diseases is preferable to trying to cure them. Rotation of crops from year to year is essential to assist in maintaining the correct physical conditions of the soil, which in turn encourages the maximum development of the plants. It is a good practice to follow Brassicas with root crops, whilst a third plot should carry such items as Peas, Beans, Celery, Lettuce, etc.

In recent years, pelleted forms of seed have been produced for the convenience of the grower. Where these are preferred, care should be taken not to bury the pellets too deeply, and it is most important to make sure that adequate water is available for breaking down the pellets should dry weather be experienced.

Having spent the whole of our lives with Suttons Seeds Ltd., we would like to express our acknowledgement and appreciation of the knowledge made available to us throughout the many years and to the assistance and encouragement given in the writing of this book. We are also grateful for the illustrations Suttons have kindly allowed us to use.

FRED POTTER
FRANK SHACKEL

METRICATION CONVERSION TABLES

The Ministry of Agriculture, Fisheries and Food has confirmed that the changeover from Imperial measure to Metric for statistics and sales in the Seed industry should be effected and seeds offered in units of milligrams, grams and kilograms. The following conversion tables will prove useful to readers.

OZ/LB to GRAMS/KILOGRAMS

$\frac{1}{32}$ oz. = 875 milligrams

$\frac{1}{16}$ oz. = 1·75 gm.

$\frac{1}{8}$ oz. = 3·50 gm.

$\frac{1}{4}$ oz. = 7·00 gm.

$\frac{1}{2}$ oz. = 14·00 gm.

$\frac{3}{4}$ oz. = 21·00 gm.

1 oz. = 28·00 gm.

2 oz. = 56·00 gm.

OZ/LB to GRAMS/KILOGRAMS

4 oz. = 113 gm.

8 oz. = 226 gm.

1 lb. = 453 gm.

7 lb. = 3·175 kg.

14 lb. = 6·35 kg.

28 lb. = 12·7 kg.

56 lb. = 25·4 kg.

112 lb. = 50·8 kg.

INCHES TO CENTIMETRES

1 in. = 2·5 cm.

2 in. = 5·1 cm.

4 in. = 10·2 cm.

6 in. = 15·2 cm.

9 in. = 22·9 cm.

12 in. = 30·5 cm.

15 in. = 38·1 cm.

18 in. = 45·7 cm.

21 in. = 53·4 cm.

24 in. = 60·9 cm.

36 in. = 91·4 cm.

ACRES TO HECTARES

1 acre = 0·405 hect.

2 acres = 0·809 hect.

3 acres = 1·214 hect.

4 acres = 1·618 hect.

5 acres = 2·024 hect.

6 acres = 2·429 hect.

7 acres = 2·833 hect.

8 acres = 3·238 hect.

9 acres = 3·642 hect.

10 acres = 4·047 hect.

20 acres = 8·09 hect.

Artichoke, Globe

Cynara scolymus

PERENNIAL

Although grown in Continental countries in considerable quantities both for home and export use, it has been rather neglected by the average grower in this country, possibly due to lack of knowledge of the subject; nevertheless it can be cultivated most successfully.

SOILS   Most types are suitable; however the best results can be found on the lighter type rather than the heavy or clay.

CULTIVATION   In the first place select a position in the garden which is sunny and well drained, this should be well manured and dug during the winter. Prior to planting the ground should also be given a dressing of a complete fertilizer at the rate of 3 oz per square yard.

VARIETIES   Although plants can be raised from seed, it is not recommended as this subject, being open pollinated, will only give a mixture of types many of which are not suitable. It is therefore advisable to purchase plants which have been vegetatively propagated, selecting either the variety Vert de Laon or Camus de Bretagne. These should be planted with 4 ft

1. *Globe Artichoke: results are more satisfactory when grown from cuttings than by raising from seed.*

15

between the rows and 3 ft from plant to plant. Plant firmly, giving an immediate mulch of short, well-made manure. These plants will produce excellent heads for a period of three years, after which they should be discarded. To maintain and, indeed, increase the stock it is advisable to take suckers from the old plants each year to commence new plantations. By this means a full season of produce will be obtained, as the oldest plants will give the first of the season's cut, whilst the second year's plants will follow, and the last planted will give the latest heads of the season.

In the colder parts of the country it is advisable to give protection during winter, using straw or bracken, removing this when the danger of severe weather has passed. It is the time to dig in a coating of manure, at the same time removing any unwanted suckers.

HARVESTING   The correct time to cut the heads is when well matured. Do not however wait for them to fully open.

CHARDS   After the heads have been harvested cut down the foliage to promote new growth. When this is about 2 ft high, tie into a close bundle, and blanch and use in a similar way to the Cardoon.

PROPAGATION   Suckers which are taken in November should be potted off and over-wintered in a frame or greenhouse, transplanting outdoors the following year when danger from frost has passed. Suckers and side growths can also be taken in April and these should also be potted up in John Innes No. 1. As soon as they are well rooted, they may be transplanted to their final quarters.

PESTS AND DISEASES   Well-grown plants are mostly trouble-free.

EXHIBITING   Seldom does one find a separate class for this subject; it does however find a useful place where there is a class for 'Any other Vegetable not listed in the Schedule'.

Artichoke, Jerusalem

Helianthus tuberosus

PERENNIAL

This useful winter vegetable is grown from tubers as opposed to the Globe type (q.v.). In the winter months of the year it is always difficult to provide that extra vegetable dish and it is therefore a subject to keep in mind when planning the various crops in the garden.

SOILS   Almost every type is suitable provided it is well drained; select a site well away from other kinds of vegetables, for the simple reason that the plants will reach a height of 10 ft or more and overshadow anything planted too close.

CULTIVATION   Provided the soil is in reasonable heart, it is advisable not to dig in manure, as this will only produce excessive foliage at the expense of tubers. However, a light dressing of a complete fertilizer at the rate of 2 oz per square yard, worked into the soil at the time of planting will be sufficient. If soil and climatic conditions are favourable, early March is the ideal time for planting in drills about 5 in. deep. Allow 3 ft between the rows, space the tubers 15 in apart in the rows, using the smaller size tubers which are preferable for multiplication purposes. Take care to see that each tuber selected for planting has at least one eye.

HARVESTING   Crop should be lifted in the autumn as soon as the tops die down, making quite certain that the ground is completely cleared, as any tubers left in the ground will commence uncontrolled growth the following year. It is a common practice to replant on the same spot in successive years.

STORING   This is quite easy. Use the same method as for potatoes; if however it is proposed to clamp outdoors it is only necessary to give a covering of straw or bracken as this subject is not affected by frost.

EXHIBITING   The tubers should be of a uniform medium size with a smooth clean skin. Points value 12.

Artichoke, Chinese

Stachys tuberifera Labiatae

PERENNIAL

This is best described as one of the lesser-known vegetables, yet those who know and grow it consider it a real table delicacy. Its small, twisted, white-skinned tubers are about the size of small prawns, and are cooked and served in a similar manner to the Jerusalem Artichoke.

SOILS   Succeeds on most types of soil which have been well cultivated and have a coating of manure, dug in during the winter months.

CULTIVATION   Open up drills about 4 in. deep with the rows 18 in. apart, spacing the tubers 12 in. in the rows. Planting time in the southern districts of the country is quite safe in early April but further north it is more advisable to wait until the last week of the month.

HARVESTING   When the foliage has completely died down in the late autumn the crop can be lifted and stored in a cool place and kept in the dark. Some growers prefer to leave the crop in the ground and lift as required for immediate table use. It is, however, advisable to lift all remaining tubers by March before they commence fresh growth.

Asparagus

Asparagus officinalis

HARDY PERENNIAL

A bed of this hardy perennial, a native of this country, is the ambition of most gardeners. Not only is it easy to grow, its produce is very much appreciated as other types of vegetables are in rather short supply during its season.

SOILS   Light soils are without question ideal for this crop. However, most other types can be made suitable provided they are well drained.

CULTIVATION   It is more important to understand that this is a crop which cannot be produced satisfactorily in a short space of time. If the grower is prepared to wait three to four years he can raise his own roots from seed, transplanting them to the prepared bed at the end of the second year, leaving the plants to become well established in the third year and cutting for the first time in the fourth year. Most growers prefer to reduce this period of time by purchasing a good stock of two-year-old roots, which you will find offered in your seed catalogue. These should be transplanted in the prepared bed at the end of March or early April. Here again in the year of planting nothing should be cut, and in the second year cutting should terminate for the season at the end of May, and in the third year and thereafter mid–end June, as this will allow the roots to build up for the following year.

Seed should be sown in late March or early April in a seed-bed, for which a piece of land well manured the previous year is most suitable. Rows should be $1\frac{1}{2}$ ft apart and the seedlings should be thinned to 6 in apart.

Preparing and planting the permanent bed   If the site has light and well drained soil the easy approach is to run single rows which should be 4 ft apart with $1\frac{1}{2}$ ft between the plants in the rows. Open up a trench 1 ft wide and about 8 in.

deep and replace a curved mound of soil about 3 in. deep in the bottom of the trench, then plant the roots. On heavy soils which may lie rather wet in winter, a raised bed should be preferred. Peg out a three-feet bed, leaving a two-feet alley either side of the bed, when digging, throw the top spit of soil from the alley on to the bed which will raise the bed to a foot higher than the alley. In using this type of bed, measure nine inches from both sides of the bed, place the line down and allow 18 in. between plants. This means that two rows are planted on the bed.

Extra care taken in planting operation is most rewarding. Open up the holes about 6–7 in. deep and form a mound; place the crown of the root on the top of the mound and cover with about 4 in. of soil. It is very important to remember to plant the roots immediately they are received and if they appear to be on the dry side do not hesitate to soak in water before planting.

General management   Keep the beds free from weeds by hand weeding and/or careful hoeing. If the summer is very hot and dry, mulch the bed with grass mowings. In November cut off the old foliage, but before doing so mark with a stick the spot where a plant has failed, which will indicate the position for planting a new root in the spring. On light soil a dressing of well-made manure can be spread over the bed; this should be left on the surface throughout the winter and lightly forked in at the end of March or early April. The crop will benefit if a complete fertilizer is applied in early March at the rate of 3 oz per square yard.

PESTS   Apart from the Asparagus Beetle, Asparagus is mostly trouble free. To overcome this trouble, should it appear, spray with Derris during the cutting season or BHC when cutting has finished.

2. *Perfection: one of the most prolific varieties of Asparagus.*

19

Beans, Broad

Vicia faba
ANNUAL

Almost any type of soil is suitable provided it is well dug and manured during the winter months for sowing in the spring; however for autumn sowing lighter types of soil are to be preferred.

Care should be taken when selecting the variety; seed catalogues indicate the various types available which show the group to which they belong, e.g. White- or Green-seeded Longpods, or White- and Green-seeded Windsors. In addition they will also list the Aquadulce Longpod which is especially recommended for autumn and early winter sowing. The Windsors are shorter-podded with a larger bean than the Longpods; the latter have a more kidney-shaped seed than the Windsors. Seed coat colour can be purely a matter of preference when the beans are cooked and served.

SEED SOWING When soil conditions are suitable in February or March, open up drills 18 in. apart and about 3 in. deep, spacing the seed 6 in. in the rows. Some growers prefer to use the double row method, in which case leave 9 in. between the rows and 2 ft between each pair of rows. Where a continuous supply is desired, it can be obtained by making a sowing in early November, using the variety Aquadulce. This should be followed by a January sowing of the same variety. To follow this, use either a Longpod or Windsor variety for March and April sowing. For the small garden where space is limited or where a picking season from June to October is appreciated, then The Sutton Dwarf Broad Bean is the answer. This can be sown in rows 15 in. apart from February until mid-July. This variety grows only 1 ft tall, and tillers out into about three or four stems which carry a prolific crop of five seeded pods which are slightly smaller-seeded than the taller forms.

This variety is also excellent for sowing under cloches or in frames, to produce an early crop.

CULTIVATION As soon as the seedlings are through the ground the use of the hoe will prove most beneficial in killing off seedling weeds and promoting good root action, so necessary with this subject as it produces its own nitrogen

3. Broad Bean: Colossal. A white-seeded variety, heavy cropper with extremely long pods of excellent table quality.

nodules. When the plants are carrying a good supply of flower, it is advisable to pinch out the top of the plant to prevent attacks in the early stages by the Black Fly. (Dolphin Fly)

PESTS   This subject is fairly free, apart from the Dolphin Fly (Black Fly), particularly from an autumn sowing, as the crop matures in most seasons before the first phase. Where spring sown crops are affected by Black Fly, spray or dust with malathion, nicotine, or derris as soon as the pest appears on the plants.

FREEZING   Most types are suitable for this purpose. However, heavy-cropping varieties with medium size pods, white-eyed at all stages of seed development, such as Meteor, have been specially bred for this purpose and can be recommended. Nevertheless, some households prefer to use the green-seeded variety Master-piece which is more attractive to the eye when cooked and served.

EXHIBITION   Always a useful dish at July and August shows, whether staged in a collection of vegetables or in the single-dish class. Amongst the many varieties available, Suttons Colossal, Imperial White, and Imperial Green Longpods are outstanding in heading the list of prizewinners. When preparing produce for the show-bench, select the specimen pods which are uniform in length and as straight as possible; avoid using any which may be showing a black hilum on the grain, for the simple reason that most experienced judges will open one pod in the dish to see the condition of the actual beans in the pod. If they are of first-class table quality, they will be free from defects such as spotted skins or imperfectly-filled pods. Maximum points value 15.

Beans, Dwarf

Phaseolus vulgaris

ANNUAL

Although during the past fifty years many improved varieties have been bred and introduced, many growers have failed to appreciate the value and importance of this vegetable. If sown in the open ground in April, it will produce table-quality pods in late June and early July, certainly a month earlier than Runner Beans. Gathered young and cooked entire, what could be more delicious? Not only do they make a change at that time of the season from Peas and Broad Beans, but they occupy very little ground for a heavy yield of produce. Furthermore, all surplus produce can be used for freezing.

SOILS   Light and well-drained soil is best. On heavier types, it is a good idea, having opened up a Celery trench, to sow Dwarf Beans on the top of this ridge; for the bean crop will have matured and be cleared in good time to commence earthing up the Celery.
Manures   Farmyard or stable manure, or compost dug in in the early winter on light and sandy soil; but on the heavier types a dressing of a well-balanced fertilizer at the rate of 3 oz per square yard will assist. Furthermore, on this type of land, a liquid feed at flowering time will result in continuous podding over a longer period.

4. *Dwarf French Bean: The Prince. The most popular variety of the English flat-podded type, continuous bearing.*

SOWING   In the southern half of the country the first sowing in the open may be made from mid-April onwards in succession, whilst further north it is advisable to wait until May. Sow a small area at intervals of a month until the middle of July to ensure continuity of cropping. Rows should be 18 inches apart. Open up a V-shaped drill about 3 in. deep and drop the seeds in at 3 in. intervals. When the seedlings are showing the first pair of leaves, single the plants to 9 in. apart.

VARIETIES   The best of the flat-podded are The Prince and Masterpiece, both being of the continuous-flowering habit. In recent years the continental pencil-podded which are completely stringless, have at long last established themselves in this country. The best of these is the variety Sprite, which is not only excellent for fresh table use but is really outstanding for freezing.

Early crops under glass   Where Dwarf Beans are required out of season, successful crops can be grown by the amateur in a heated greenhouse, either in a border or in nine-inch pots. Sowing distance in the border is the same as for outdoor cultivation; in pots, sow three seeds to each pot, using J.1.3 compost. Maintain a temperature of about 60°F. Commence liquid feeding when the plants have made good root growth.

Waxpod   this is a golden-podded form of the dwarf bean which is considered by many to be a delicacy. Kinghorn Waxpod which is a stringless variety has now superseded the old variety Mont d'or and cultivation is identical outdoors to normal varieties.

Haricot flageolets   This is a term used in France where the seeds are used in a state intermediate between green and ripe, removing the beans from the pod and cooking immediately. The most popular variety for this continental dish is Chevrier Vert; cultivation the same as for normal English types.

Haricot secs   This is a continental name for dried haricot beans for winter use. A number of varieties are suitable for this purpose; however, where a smaller white round seed is preferred, the variety should be Comtesse de Chambord. This has a very bushy growth and carries a heavy crop of short pods. Seed should be sown from the middle to the end of April, in rows 20 in. apart and 12 in. between the plants in the rows. Allow the pods to ripen fully on the plant, after which they can be picked and shelled; or pull the plants and thrash out the beans with a stout stick. After cleaning, store them in a dry place.

EXHIBITION   When harvesting pods for exhibition, select young straight pods of uniform size and colour, about 5 in. in length, with the girth of a pocket pencil. Points value 15.

Beans, Climbing French

Phaseolus vulgaris
ANNUAL

A good description of this subject would be that it is a climbing form of the Dwarf French Bean, growing to about 5 ft tall. It is, therefore, advisable to support with bushy pea-sticks.

SOILS   Most types are suitable and should be dug well in advance of the sowing season. A dressing of farmyard manure or compost should be given; do not, however, overdo the feeding as this only leads to a lush plant growth at the expense of pod production.

SOWING   The end of April and throughout May is the most suitable period, in rows 3 ft apart, using a V-shaped drill about 2 in. deep, spacing the seeds about 3 in. apart. As soon as the plants show the first pair of leaves, thin to a distance of 6 in.

VARIETIES   One of the most popular is Earliest of All, the pods of which very closely resemble a pencil-podded stringless dwarf bean. Pods may be gathered and used fresh, but, being white-seeded, can also be allowed to ripen and used as white haricots in the dried state for the winter.

In recent years the Pea Bean (Coco Bi-colour) of French origin has become popular in this country. Pods of this are short and broad and, gathered young, are really excellent when cooked whole. This variety can also be left on the plants to ripen and used as a haricot in winter, having a pebble-shaped seed. Another rather unusual, yet attractive, variety is the Purple Podded, sometimes better known under the name of Purple Caseknife or Indigo Blue. When the pods are fit to gather for fresh table use, they are completely purple.

Ornamental Podded   As its name suggests, the pod when fit to gather has a golden background with scarlet markings which make its appearance most attractive; hence the reason why this is grown and staged in vegetable groups at shows. For the record, this variety was also known years ago under the name of Firefly and also Robin's Egg.

Beans, Runner

Phaseolus multiflorus
ANNUAL

This vegetable is mostly used and considered as as an annual. However, by saving the roots from year to year (not to be recommended) it can be classified as perennial. Although it is recorded that the Runner Bean was first introduced into this country in 1633 from South America, its development and uses during the next three centuries were very slow; nevertheless, few, if any, would disagree that today it is one of the most popular of vegetables. In fact, it is almost exclusively British and is seldom seen growing in other countries.

SOILS Whilst this subject can be grown on most types of soil, it prefers the lighter type. The plant possesses a very extensive root system; deep digging and liberal manuring are therefore essential. The land should be prepared during the winter months, preferably not later than March.

A simple approach is to remove the top soil to a depth of 15 in. where possible, making the trench 2 ft wide, and, before placing in the manure or compost, thoroughly break up the sub-soil, restoring the top soil to about 2 in. of the normal ground level.

Provided the soil is in a reasonable physical condition, and having applied a good coating of manure or compost in the base, take care not to supplement this with an artificial fertilizer with a high nitrogen content, as this could lead to an extremely lush growth of foliage, at the expense of both flower and crop, and could well contribute to bloom-drop, so often experienced in many areas. Bloom-drop can be counteracted by dressing the crop with sulphate of potash at the rate of $\frac{1}{2}$ oz per yard run immediately after watering. In many cases the cause of poor pod-setting in some districts has proved to be the scarcity of pollinating insects, the most efficient of which is the bumble bee, and it is quite a common practice to place a saucer of sugar water close to the beans to attract bees.

TIME OF SOWING In the more southerly parts of the country sowings outdoors may commence in early May; in more northerly districts where late May and early June frosts are possible, delay sowing until the end of May. Many growers sow early to mid-April where cloche protection can be given, removing the cloches when danger from frosts has gone. Those gardeners with polythene tunnels may sow about mid-April.

Where the cold greenhouse or cold frame is available, sowings can be made in boxes or trays in the latter half of April, using John Innes No. 1 Seed Sowing Compost, Levington, or other forms of Peat Compost for filling the trays. When raising plants by this method the utmost care must be taken to harden the plants before transplanting, as failure to do so often results in a chilled plant which retards growth and in some cases causes complete failure.

GROWING METHODS If the trench has been prepared as mentioned above, the best method is to run a double row in the 2 ft wide trench, using bean poles, canes, or stakes with a final spacing of 9–12 in. between the plants, sowing the seeds at a depth of 2–3 in. at two seeds per pole or cane.

Alternatively, in the same prepared trench two good stout poles or posts could be erected

FIGURE 1 *Runner bean support, normal pole or cane system*

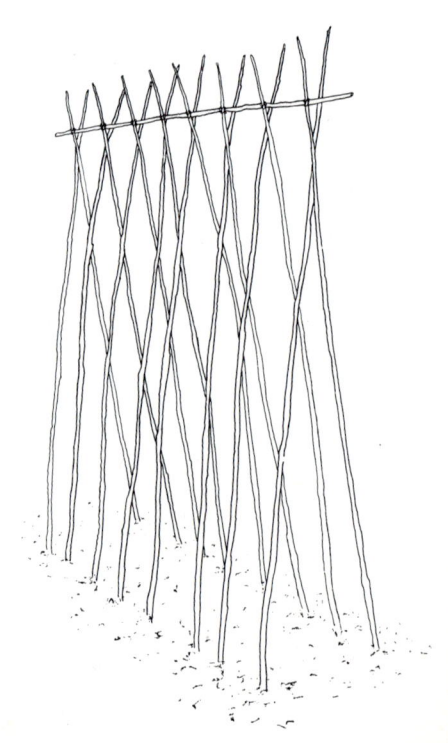

with galvanized straining wires placed at the top of the poles, a second straining wire at about 4½ ft, and a third 6 in above ground level (see drawing). Fillis string or binder twine can then be cut to the required lengths, tying to the top wire, looping it round the second wire, and fastening it to the bottom wire. This is a less expensive method than the use of bean-sticks or canes; further, there is no reason why this structure cannot be left standing for a year or two—thus saving work.

Another method which can be very usefully employed, particularly in the smaller garden, is to erect a good stout centre pole, fixing Fillis or binder twine to the top. Run the strings down, anchoring to the ground by pegs, thus forming the shape of a bell tent or maypole (see drawing). This will enable a quick-growing salad crop to be grown on the ground around the main pole, which will have matured before the beans have climbed and excluded the light.

Where the early production of Runner Beans is desired, sow in rows 2 ft apart and, as soon as the climbers begin to appear, immediately pinch them out to form a bush type of plant which does not require support. For this form of cultivation it is essential to use a suitable variety such as Kelvedon Marvel or Sunset, which are early-flowering and close-jointed. An alternative to this method is to use the Hammond's Dwarf Scarlet Runner which requires neither staking nor pinching and, when fully grown,

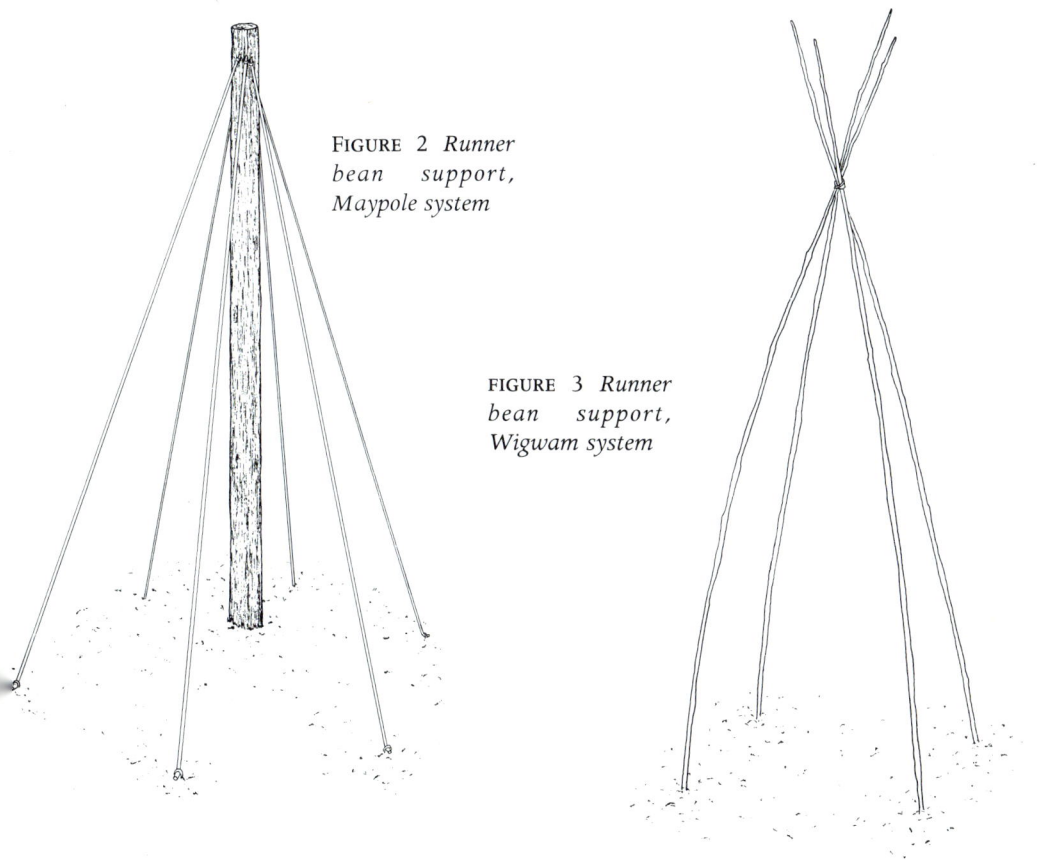

FIGURE 2 *Runner bean support, Maypole system*

FIGURE 3 *Runner bean support, Wigwam system*

5. Runner Bean: Enorma. Smooth podded with extreme length, first-class for exhibition and table use. Excellent for freezing.

be lifted at the end of the growing season, stored throughout the winter and replanted the following year. It is true that this can be done, if stored in a frost-proof building in peat or sand; however, years of experience clearly demonstrate that it is not to be recommended, for the simple reason that the cropping is much below that of freshly-sown seed, and it is the heaviest weight of pods from a given area which is required.

GENERAL CULTIVATION   In the early stages of growth, particularly on heavy soils where drainage is slow, slugs can be troublesome; therefore precaution should be taken by the use of a proprietary brand of slug-killer.

Under drought conditions the plants will appreciate copious watering, especially on light soil, but as with all other subjects the use of the hoe from time to time will not only destroy weeds but the tilth created will assist in conserving moisture. Many experienced growers will, in the event of dry weather at flowering time, spray with soft water during the evening, which will assist the setting of the flowers and will ultimately prolong the season of bearing—provided the pods are gathered at regular intervals, as nothing will stop a plant bearing more quickly than when beans are left to develop seeds in the pod.

Use a B.H.C. insecticide to control Black Fly if the Runner Beans are attacked.

FREEZING   Surplus produce after table requirements have been met can be used for freezing and Beans respond very well to this method of processing for out of season use. If growing for this purpose only, the varieties Achievement (scarlet flower) or White Achievement (white flower) have been proved most suitable, but, provided the pods are gathered young, most varieties may be used.

EXHIBITION   Reference has already been made to the popularity of this subject for table

has a maximum height of about 15 inches. Hammond's Dwarf Scarlet is extremely early-cropping and requires constant picking to encourage continuity of crop.

On no account should the seed be soaked in water prior to sowing as this method has been found to encourage Halo Blight, which can be a very troublesome Bean disease that spreads rapidly under adverse growing conditions and can result in a very disappointing crop.

Frequently the question is asked can the roots

use, but it is equally popular at shows. Many dishes exhibited have been grown for the dual purpose of both table and exhibition. Where the ambition is to grow the longest pods with first-class quality, spacing of the plants should be increased to 18 in. from plant to plant.

Sowing-time depends entirely upon the date of the show; for example, for late July to early September, sowing dates would be during May, according to district. In colder regions where growth is slower in the early stages, it is advantageous to sow in boxes under glass, transferring the plants to the open ground when ready and when the weather is suitable. On the other hand, for late September/October shows, sowing should be delayed until the second or third week in June, according to the district.

If by chance, as can happen, the plants, when, say, 3–3½ ft tall, indicate by the size and colour of foliage that a tonic is necessary, a liquid feed or a top dressing with a complete artificial manure, hoed and watered in, will restore vigour and assist growth at a critical period.

To obtain the maximum length and quality of pod; when it can be seen that three or four flowers on the bloom spike have set, leave the two most advanced small pods, removing all others on that particular spike. This will throw all the effort of the plant into producing the necessary requirements.

The question is frequently asked how long it takes from the time the flower sets until the pods are ready for the show-bench. This is a difficult question to answer, due to varying climatic conditions, but the average would be somewhere about 18 days.

Expert growers will run their fingers down the pods and apply slight pressure where the little seeds are forming to prevent development, thus rendering them free from any trace of 'beaniness'.

The hallmark of the first-class exhibitor of Runner Beans is indicated by a number of factors which are immediately recognized by show judges.

Firstly, when harvesting the beans, cut completely from the stem of the bloom spike, leaving a nice cross-section on the end of the bean stalk.

Secondly, if possible, cut the beans the night before the show, roll them singly into a wet tea towel until the whole bundle is rolled completely. This will assist in having pods completely straight and in the freshest condition.

If necessary, cutting off the pods may commence a few days before show day, and the beans may be kept stalk downwards in a container with about ½ in. of water, which should be changed each day.

Thirdly, when selecting produce, reject any pod which appears to be pale green in colour, any which have commenced to show the formation of seeds, or those which have bottle-shaped necks (stalk end). Uniformity of each characteristic of the exhibit is essential. Maximum points value 18, as outlined by the Royal Horticultural Society Rules for Exhibiting and Judging.

Beet

Beta vulgaris

BIENNIAL

Although the wild form (*Beta maritima*) can be found growing around our coast, it was not until the fifteenth century that the cultivated forms were used in this country. Most varieties used up to about sixty years ago were of the long-rooted shape, but today it would be quite true to say that the only remaining long variety would be Cheltenham Greentop, which is grown by market-growers and amateurs for storing and winter use.

6. One of the globe beet series, the Boltardy is less prone to bolting than most beets and suitable for early spring sowing.

One might well ask why the change to the globed-shaped varieties; the answer is that they are much quicker-growing, lend themselves far better to successional sowings, are much superior in colour and texture of flesh, and are easier for the housewife to cook than the old-fashioned long types.

Thanks to modern plant breeding, various selections of the Globes are available wherein each type serves a special purpose. For example, if the old type of Globe has been sown in March, many of the plants would have bolted to seed, but today the varieties Boltardy and Early Bunch are much less prone to this trouble when sown early. Furthermore, produce is ready for the table from June onwards. The long types grown in the past occupied the ground for a whole season, whilst the modern approach is to sow little and often. Therefore, having made a March sowing for June/July use we can follow this with another in May to form a succession of young, table-quality produce. The last sowing of the year should be made in July, having cleared a piece of ground which has already produced early Potatoes, early Peas or autumn-sown Broad Beans. For this sowing, do not dig the ground, as this will result in a loss of moisture. Clear the debris, create a good tilth by using the hoe, draw the drills, and, if the ground is very dry, run the water-pot up the drills, leaving them for an hour before sowing.

In the more southerly half of the country the last summer sowing should be made about mid-July, but further north sowing should be advanced to the first week of July. In the more favourable parts of the country the roots from this last sowing may be left in the ground and used as required; in the event of a cold spell during the winter a covering of straw or bracken will afford protection.

In some exposed areas and on some soils, it would be advantageous to grow a crop of the Cheltenham Greentop as a maincrop for storing for winter use in addition to a crop of the Globe type for early use.

SOILS   Light and medium loams are preferable for the early-maturing Globe varieties; the heavier soils are suitable for the Cheltenham Greentop. Do not manure the ground at the season of sowing; use ground which had been manured during the previous year. If, however, it is considered that the soil is below standard, or not in a reasonable state of fertility, give a dressing of a complete fertilizer before sowing at the rate of 3 oz. per square yard, working it into the ground with the hoe.

CULTIVATION   For all Globe varieties drills should be 12 in. apart and $1\frac{1}{2}$–2 in. deep. When the seedlings show the third leaf they should be thinned out to 3–4 in. Cheltenham Greentop is best sown in early May with the rows 15 in. apart, and a final thinning distance of 8 in. Lift the roots in October and store either in a clamp, well covered by straw, or in boxes of sand, peat, etc.

EXHIBITION   Care should be taken in reading the schedule of the show as in some instances it will read 'complete with foliage', but sometimes it will be found to read 'with not more than

three inches of foliage'. When selecting the required number of specimens, maintain uniform size and shape, and avoid coarse roots. Wash them in cold water with a soft sponge; do not use a brush as this will scratch the skin. Watch carefully the base colour of the root after washing; a good dark base will indicate a good flesh colour. Avoid anything pale, as this may be the very root the judges will cut on the show bench.

PESTS AND DISEASES   Very little trouble will be experienced in this direction, but one disease which can be common in this subject is Scab. This is due to lack of humus in the soil together with the excessive use of lime.

LEAF BEETS   Under this heading one must include Silver or Sea Kale Beet (Swiss Chard). It does not have an edible root but is grown for three distinct reasons. It is ornamental when growing; its broad white mid-rib is used as a substitute for Sea Kale; while the foliage can be used as Spinach. Another Leaf Beet is the Perpetual Spinach; for full details see Spinach.

Rhubarb Beet is similar to Swiss Chard in all respects except for the fact that the mid-rib is red, not white. In the cultivation of these leaf types, seed should be sown in early May in rows 18 in. apart and singled out to 12 in. between the plants.

Borecole

Brassica oleracea acephala

B. napus

BIENNIAL

Its common name is KALE, sometimes referred to as Scotch Kale. This, however, can be misleading as, in addition to the curled-leaved Scotch varieties, there are the plain-leaved varieties, such as Cottagers, for human consumption; for feeding dairy herds the Marrow Stem; and for sheep feed the well known Thousand-Headed; and all of these belong to the same family (*Brassica oleracea acephala*).

Also under the heading of Kales is another section, but from another side of the brassica family, which are best described here as Rape Kales. Included in this section one would find the varieties Asparagus, Hungry Gap, Ragged Jack, and Russian, for many of which it would be difficult to find seed today, apart from Hungry Gap. This subject can, therefore, be looked upon as a form of insurance, being one of the hardiest of the brassica family.

CULTIVATION   To grow a good crop of the Curled Scotch or Cottagers, sow in a seed-bed in May, transplanting to final quarters in July in rows 20 in. apart with 18 in. between the plants. As this crop is to stand the winter, select a portion of the garden which is well drained. A useful method is to manure for early Potatoes, transplanting the Kale after lifting the Potatoes. When the Potatoes are lifted, give a dressing of a complete fertilizer (2 oz per square yard), and work this into the surface, consolidating the soil as quickly as possible to conserve moisture. When the plants are established run the hoe through the soil to destroy any weed seedlings. Rape Kales, previously mentioned, must at all times be sown *in situ* and thinned down. Best results will be obtained by sowing in shallow drills in July, rows 15 in. apart, and singled out to 12 in. between plants. It is important to remember that this subject should never be transplanted. Here again it can be recommended to sow this after early Potatoes have been lifted, or early Peas cleared; treat the Kale as a 'catch crop'. It is extremely hardy and remains fit for use long after other forms of brassica have run to seed.

PESTS AND DISEASES   From a disease point of view seldom does one experience any trouble except on land which has club root. Protect the seedlings from flea beetle as they emerge through the soil by using Sevin Dust.

Broccoli, Sprouting

Brassica oleracea botrytis

BIENNIAL

The student of history would find that the original name for Broccoli was applied only to the sprouting types. For many years the winter- and spring-heading varieties were also included under the name of Broccoli, but in the last few years this has been corrected. Most seedsmen now list what was formerly under this name in the Cauliflower section, so see Cauliflower for all single-heading varieties.

Few gardens are without Sprouting Broccoli, as it could be safely said that it is a popular British dish, easy to grow, the Purple and White Sprouting being available when other forms of the brassica family are in short supply. The Italian Green Sprouting, known as Calabrese, which is ready for cutting from September until the weather closes down about November, is not only a delightful dish cooked fresh, but is also excellent for freezing.

CULTIVATION Like all members of the brassica family, rotation of crop is essential. Avoid using for the Purple and White forms, which stand through the winter, land which is not well drained, as the plants can be badly affected and may die if their feet are standing in water. Dig the land during the winter months, having applied a coating of manure or compost, thus giving the land plenty of time to consolidate.

Growers generally sow in a seed-bed using a V-shaped drill about one inch deep about mid-May. Sow thinly as this will assist in producing a good, stocky, well-grown plant with a healthy root action. A thickly-sown seed-bed will only produce long drawn-up specimens which seldom make first-class plants. Transplanting should take place about mid-July in rows 20 in. apart with 15 in. between the plants.

Purple Sprouting can also be sown *in situ* and thinned down; if this method is used, delay sowing until June, using the same distance between the rows; but a foot between the plants is sufficient. It is not necessary to produce thick stems, as it is the head of the plant which will produce the sprouting growth. Do not overdo the use of nitrogen as plants which have been overfed will produce only a soft growth which is more easily affected by frosty conditions.

To produce a crop of the Italian Green Sprouting (Calabrese) it is essential to sow in a seed-bed at the end of April or, at the latest,

7. *Broccoli: Purple Sprouting. Extremely hardy for cropping from Christmas till April.*

8. *Brussels Sprout: Early Button. An early type with a walnut sized deep green button.*

early May, transplanting as soon as the plants are large enough, usually about the middle of June to mid-July according to district. The rows should be 20 in. apart, with 18 in. between the plants. The plants will benefit if the centre head is cut immediately it is ready as this will enable the plant to develop the remainder of the sprouting growth.

VARIETIES Three distinct strains of Purple Sprouting are available—Early which is ready for use in January/Febraury and can be recommended only for the more southerly parts of the country: Sutton's Purple Sprouting which is probably the best for all districts, maturing in March/April, and is less affected by frost than the early strain: the latest variety is listed under the name of Late April Purple Sprouting; it is more dwarf than the two previously mentioned. In the smaller type of garden, some growers consider this to be rather too late in maturity to allow the ground to be cultivated in time for a further cropping programme.

Improved White Sprouting can be recommended; this matures in March/April, the Late White Sprouting being fit in Late April. Here again the same remarks apply as to the Late April Purple Sprouting.

PESTS AND DISEASES In general terms, few should be experienced, but it is useful to remember to watch for flea beetles on the young seedlings, and to prevent aphids, which can carry virus, by spraying.

Brussels Sprouts

Brassica oleracea bullata

gemmifera

BIENNIAL

Developed from the Wild Cabbage, its origin is probably Belgium, from whence it derived its name. Nevertheless, it has established itself in almost every private vegetable garden in this country, and vast acreages are produced by commercial growers, not only for fresh table use but also for freezing, thus making this subject available for most of the year.

Thanks to a very extensive breeding programme over the past half century, few vegetables have undergone so many rapid changes. As a result many of the old varieties are no longer available and in their place we have far superior types both in 'straight' varieties and F_1 Hybrids. Indeed, the fashion has changed almost completely. The requirements of the present-day consumer are usually for walnut-size buttons as against the large size of the past; and there are degrees of maturity—Early, Mid-season, and Late, and spacing of the buttons to prevent rotting on the stem and for ease of gathering. All in all a great deal of pleasure and pride can be found in growing this crop.

CULTIVATION Heavier soils are most suitable, which means that the ground should be prepared as early as possible in the winter to allow the frost to break down the soil. On light land there is the same need for early winter preparation; in this case, however, it is to allow the soil to consolidate in readiness for planting. Loose soil is anything but ideal and can result in 'blown' or loose buttons and, in exposed places, the plants may be blown over by the wind.

Sprouts are gross feeders, due to the long season of growth, so a good coating of farmyard or stable manure is required, or, failing this use as much compost as possible. If in any doubt concerning the physical condition of the plot, do not hesitate to give a dressing of 3 oz per square yard of a well-balanced fertilizer, which should be applied and hoed in just prior to transplanting.

Where produce is required for September/October gathering, it is essential to make an early sowing under glass at the end of February or early March. In the greenhouse the best approach is to fill a pot, say 48 size, with John Innes Seed Sowing Compost, broadcast the seeds over the surface, and lightly cover with the same soil compost. As soon as the seedlings show their first pair of leaves, prick them out into seed-trays, bringing them along in the greenhouse until they are ready for transplanting to their final quarters outdoors. Special care should be taken that before transplanting, the trays are stood out in a sheltered spot outdoors for the plants to become well hardened before transplanting. It should be clearly understood that any biennial subject raised under glass which has not been carefully hardened will be more prone to the risk of bolting due to vernalization, i.e. warm days followed by cold nights, which chills the plants, and they endeavour to complete their life-cycle by running to seed. For those who do not possess a greenhouse, the same procedure can be followed by using a cold frame; but to save labour it is also feasible to draw drills in the frame, sowing the seed *in situ* in a similar manner to raising plants in the open ground.

In the more southerly parts of the country outdoor sowings about mid-March in a seed-bed may also be made. Where produce is not required at an early date but is appreciated from November to February, sowings may be delayed in a seed-bed outdoors until April, according to the geographical position of the district.

As already mentioned, this subject requires a long season of growth; therefore, as soon as the plants are large enough, they should be transplanted to their final quarters—by the end of May or early June at the latest—giving a minimum distance of $2\frac{1}{2}$ ft square.

VARIETIES   Reference has already been made to the importance of the introduction of F_1 Hybrids. These can be recommended in preference to many of the old varieties, except where really late produce is essential. Most of the present-day F_1 Hybrids such as Peer Gynt serve as a good general purpose crop, say, for October to early January, but where Sprouts are appreciated into February and March, then the blue-leaved Bedfordshire types, such as Winter Harvest and Market Rearguard are outstanding for their main crop and long-standing character.

FREEZING   If this is undertaken, the F_1 Hybrids are preferable for the purpose, as they are the correct size, solid, and with an attractive colour. The uniformity of size of button is an added feature of these hybrids.

The grower can to some degree control the time of maturity of the Sprouts by a very simple operation. If the plant has made average growth by, say, early October, with the Sprouts beginning to show, he can advance final maturity by rubbing out with the thumb the terminal bud in the head of the plant. To make this operation abundantly clear, assume that 40 plants have been put out. If 20 plants had terminal buds rubbed out, these would mature much earlier than the 20 plants left in their natural state. It will be seen, therefore, that, where the grower is using one variety, he can extend the season of use by this simple operation.

EXHIBITION   It is only in late September/October shows that this subject is exhibited. For a prize-winning dish great care should be taken to maintain uniformity of size and solidity; and above all avoid stripping off several layers of leaves, as each layer removed reduces the colour from a dark to a very pale green, insipid-looking Sprout—something experienced judges will not tolerate. Points value 15.

PESTS AND DISEASES   To guard against Flea Beetle, dust with Sevin Dust outdoor seed-beds immediately emergence takes place. Prevent Aphid attacks by spraying with Malathion.

In some wetter parts of the country Ring Spot can be troublesome and is usually found where crops have been heavily manured, thus giving a soft growth, and in many cases when planted too closely. Some measure of control can be brought about by burning Brussels Sprout refuse, and also by practising a rotation of crop where this is possible.

As with other members of the Brassica family, Club Root can attack this crop, and the use of Calomel Dust prior to sowing in the seed-bed and also when planting out is an essential operation.

White Blister, which is a fungus disease, is not really important, apart from making the plants unsightly. Any affected leaves should be removed to encourage better circulation of air around the plants.

Cabbage

Brassica oleracea capitata

BIENNIAL

This extremely interesting and valued member of the brassica family plays an important part in our everyday requirements of fresh vegetables. In general terms it would be safe to say there are very few weeks in the year when these are not available, for, not only do they provide an alternative to other members of the brassica family, but they are reasonably simple to grow. The wide range of varieties which is available stresses the importance of the utmost care in selecting the correct variety for the desired season of use. For example, some varieties will mature in about 12 weeks from the time of sowing, whilst others require a full season of growth. And there are various shapes, such as pointed-headed, ball-head, and drumhead. Some varieties are suitable for sowing under glass for transplanting into the open ground, others for spring sowing in a seed-bed outdoors for summer, autumn and winter maturity.

Added to all these are a number of choice varieties specially bred for autumn sowing, which are less prone to bolting in the spring, and these are fit for cutting in April and May. Colour of the leaf also plays a part—there are the typical English pointed-headed varieties which give that beautiful dark green colour when cooked; the Scandinavian ball-head types are more white than green after cooking; and, of course, there is the Red Cabbage which is so popular for pickling, but can also be cooked in the ordinary manner.

CULTIVATION   As with all Brassicas, Cabbage should be grown on land which is rotated and following such items as Peas, Beans, Onions, and, indeed, Potatoes.

Preparation of the land should follow as for other Brassicas, digging it as early as possible and incorporating either a good coating of manure or compost, supplemented where necessary with a dressing of a complete organic fertilizer of 2–3 oz to the square yard prior to transplanting. Having said this, it should be pointed out that this recommendation applies particularly on land which is used for transplanting a late January or early February sowing under glass, eventually transplanted outdoors in April for maturity in June; or for plants raised in a seed-bed in April outdoors and transplanted in June for late summer use. For main-crop types of Cabbage for winter use such as January King it is the general practice to transplant these after lifting early Potatoes or clearing early Peas or autumn-sown Broad Beans; if the soil has been well prepared for Potatoes the residue of the manure should still be available. It is advisable, however, to give a light dressing of organic fertilizer just prior to transplanting.

Autumn-sown varieties for spring cutting will succeed best if they follow the Onion bed. Having harvested the Onions in September, clean off the ground by the use of the hoe and transplant without further manuring. If the Onion bed was well prepared there will be sufficient food left in the ground to allow a little growth between the time of transplanting (end of September/early October) and the weather closing down. As soon as weather conditions permit in early March, then is the time to top dress. Some growers prefer a complete, well

9. *Cabbage: January King. A form of insurance against winter conditions, being the latest, hardiest and most long lasting.*

balanced fertilizer such as Elliotts or Growmore, some prefer Sulphate of Ammonia, but if extra quick growth is required Nitro-Chalk can be used. But where this is used, you will find that, due to the soft lush growth, the produce quickly flags after cutting; such produce should, therefore, be cut and used the same day.

VARIETIES AND THEIR USES   For the convenience of quick reference in selecting the correct variety for its specific purpose, it is necessary to divide this subject into seasonal sowing and transplanting dates. If you have a greenhouse, sow seed of the variety May Express in a 4- or 5-in. pot the last week of January in a temperature of 50°F, pricking the seedlings out into seed-trays which have been filled with J.I.3 or a similar compost. As soon as weather conditions permit in early April, the tray should be stood outside in a sheltered spot for a few days to harden the plants thoroughly before transplanting into the open ground, allowing 20 in. from row to row and 15 in. between plants. This will give a good supply of ball-head Cabbage in June when other vegetables are in comparatively short supply. If, however, a pointed-head variety is preferred use Sutton's Earliest adopting the same methods as for May Express.

Another ball-head variety which can be sown and grown on similar lines is June Star, which will mature after May Express and is quite a useful Cabbage, not only for its table qualities, but also where produce is required for July shows.

The following varieties, sown in a seed-bed in April, transplanted in June, would be fit to cut—Pride of the Market (ball-head) or Sutton's Earliest (pointed-head) for August; Autumn Supreme (semi-ball-head) for September/October; Christmas Drumhead October/November; January King (drumhead) December/March.

Sow in early August in a seed-bed, transplant to final quarters end of September or early

10. *Cabbage: May Express. A quick growing round headed summer Cabbage, ideally suited to sowing under glass at the end of January–February for June cutting.*

October: Sutton's April for April/May cutting, but, where a larger head is required, slightly later in maturing, use Flower of Spring, which heads in May. Both these varieties are pointed-headed, with a rich dark green colour.

It will be seen that the above recommendations, if used in their entirety, would provide a continuous supply of Cabbage throughout the whole year.

Collards (Greens)   Although commercial growers produce this item during a large part of the year, very few amateur growers realize the true value of this crop. It is simple to grow and occupies the ground for a short space of time; yet the produce provides a delicious dish, particularly in the southern half of England from November to March or April. Seed can be sown about the last week of July in drills 1 ft apart, using the variety Early Giant, and, as soon as the seedlings are showing the third leaf, they should be singled to 3 in. apart in the rows. The plants will make a leafy, non-hearting growth, and, as already mentioned, will be ready for cutting as Greens from November onwards. To assist quick growth and to develop a nice dark green colour, a little Sulphate of Ammonia run down by the side of the rows and hoed in is most beneficial.

Winter Stored Cabbage　In Northern European and Scandinavian countries it is necessary to grow the large white ball-head Cabbages, which are harvested in the late autumn and stored for winter use. This type of Cabbage is available for growing in this country, and large acreages are now being grown commercially. The variety used is Sutton's Winter Salad. Seed is sown in a seed-bed in April and transplanted in June, allowing 20 in. from row to row and 18 in. between the plants. In November the heads should be cut and all outside leaves completely trimmed away from the head, leaving the white solid head only for storing. Two methods may be used. Firstly they may be stored in a clamp outdoors, which is accomplished by placing a good deep layer of straw on the ground, putting a very thick protective cover of straw over the whole of the heap, and inserting into the top of the clamp either an agricultural drain-pipe or a good twist of straw, to form a vent. The whole clamp should then be covered with a coating of soil to hold the straw in place. Alternatively, if a cool building is available, place some slats on the floor, put the Cabbages on the slats in a heap, and cover with straw. It is advisable from time to time to open up the clamp or heap and

examine the heads, trimming away any infected foliage, to maintain a clean healthy head. This practice could be used to advantage in the more northerly and exposed districts of the country.

EXHIBITION　It is always a debatable item amongst exhibitors as to which is the most suitable variety, due to the fact that shows can cover a period from June to November. Much, therefore, depends on the date of the show. One thing that is quite certain, after years of experience in exhibiting, is that the most popular exhibition variety is Winnigstadt. This is a neat, pointed-headed variety with a solid heart; if the shows are being held in August to October, this is undoubtedly the variety to use. One must, however, pay special attention to sowing and planting times to fit the date of the show.

The ball-headed types, such as Pride of the Market and Primo can be grown and used quite successfully for July and August shows. Maximum points value, 15.

Ornamental Cabbage　Although these are not generally grown for edible purposes, with the ever increasing interest in floral art they are fast gaining popularity for this purpose. A packet of 'mixed' seed, will contain four distinct types—plain and frilled Green-Leaved, and plain and frilled Pink-Leaved. Quite apart from floral art, they are extremely decorative when grown in a flower border.

Seed should be sown in a seed-bed in April or May, transplanted June/July 18 in. apart and 15 in. between the plants, for maturing in the late autumn and early winter.

The matured plants have a loose leaf formation, but manuring and cultivation should follow the recommended lines for general Cabbage growing.

Portuguese Cabbage　In the early part of this century this was a very popular item in the kitchen garden of the old estates. Unfor-

11. *Cabbage: April. An outstanding 'Spring' Cabbage, small, compact habit. Dark green in colour with rapid development.*

tunately, however, in recent years it has seldom been seen. Nevertheless, it is a most delicious vegetable, non-heading, but producing large leaves with a broad mid-rib, which when cooked is like Sea Kale, whilst the foliage may be cooked in a similar manner to ordinary Cabbage. Sow in a seed-bed April/May, transplant in June/July, ready for use late September to December. A well-prepared soil as for ordinary Cabbage is essential. Allow 2 ft from row to row, 20 in. from plant to plant. Also known as Couve Tronchuda.

PESTS AND DISEASES Take preventive measures against Flea Beetle as the young seedlings emerge through the ground, or treat the seeds with Murphy Combined Seed Dressing prior to sowing.

Leaf Spot and Ring Spot may be experienced in some districts. It can be caused by soft growth, encouraged by heavy manuring and too much nitrogen.

Mosaic (Virus) is caused by Aphid attacks, which can easily be prevented by spraying with a Systemic Insecticide to kill off these insects. Probably the most common trouble is Club Root, which is dealt with in another chapter (see p. 35).

Cabbage, Savoy

Brassica oleracea bullata

BIENNIAL

There are a number of varieties available which, if they were all sown together in a seed-bed outdoors during April/May and transplanted late June/July, would form a natural succession for cutting from August to March or April. The early-maturing varieties such as Best of All are excellent for exhibition purposes in August/September shows, but the true value of the Savoy is in its hardy character to withstand winter conditions, when varieties such as Winter King and Rearguard are most appreciated.

It should be remembered that a well-hearted Savoy is most valuable in January to March when most other fresh vegetables are in short supply.

Preparation and manuring of the land is the same as for spring-sown Cabbage; seed should be sown in a seed-bed (for autumn cutting varieties) in April, transplanted in June, 20 in. between rows and 15 in. in the rows.

For winter cutting, sow in seed-bed early May, transplant in July, using the same spacing.

Cabbage, Chinese

Brassica cernua

ANNUAL

Although this is a member of the Brassica family it is quite distinct in appearance and in cultivation from ordinary Cabbages. One could not do better than to describe it, when fully matured, as being similar in formation and appearance to a larger form of Cos Lettuce.

In this country its popularity is increasing rapidly, and it is extensively grown not only in Asia but also in America. At one time the Pe-Tsai or Michili were the two varieties mostly grown in Europe, but in the last two or three years new F_1 Hybrids such as Sampan are being grown in this country, being entirely self folding with a

12. Chinese Cabbage: Sampan F_1 Hybrid. Easy and quick to grow, excellent cooked or raw in salads.

good tight solid heart and less prone to bolting than the old varieties.

Several good cooking recipes are available; the two most popular are boiled and served plain, or cooked with a quantity of shrimps. It also makes a very interesting and tasty addition to salads.

CULTIVATION This crop can be grown on most types of soil which are in reasonably good heart, in fact land which will grow a good lettuce crop is ideal for this subject. To succeed it is important to remember several factors. Don't sow outdoors before June; but sowings can be made any time from early June to the middle of July. Secondly, sowings must be made *in situ* in rows 15 in. apart, thinning the plants to a 12 in. spacing in the rows. Thirdly, in no circumstances must transplanting take place. The reason why earlier sowings than those recommended above should not be made is that, when sown earlier, plants will bolt to seed without making any useful edible material. The plant is very quick growing, taking approximately 9–10 weeks from date of sowing to full maturity. In the event of dry weather this plant responds to water.

With the July sowings, if a dry period is being experienced, open up the drill, give this a good watering and leave for an hour or so before sowing the seed. This will assist quick germination.

Cape Gooseberry

Physalis edulis

PERENNIAL

The Cape Gooseberry is a tender plant and requires greenhouse treatment in this country. Seed is sown in February and March in a temperature of 65–70°F.

It is also possible to propagate by cuttings of half-ripened wood, taken between January and April and rooted in sandy soil in a temperature of 65–70°F.

The seedlings should be pricked out and finally potted singly into 5- or 6-in. pots. A fairly rich compost is necessary, and the pots should be placed in a sunny position. The plants require frequent watering, and a weekly application of a liquid manure between May and September is helpful. Fruits should be gathered when ripe and fully coloured in September.

Very decorative and delicious taste, raw or sugared. It is also possible to plant out in a warm sheltered border.

Capsicum and Chili

Capsicum annum C. baccatum
ANNUAL/PERENNIAL

Under cultivation, especially the type used in this country, this plant is an annual, but in warmer climates several species may be perennial. The growing of this group is a fairly simple procedure, as the seeds can be sown thinly in pots or pans of fine soil in February or March under glass where a temperature of 55°F can be maintained. Pot on the young seedlings singly into 3-in. pots as they develop and try to keep them growing without any check, keeping them in heat until well rooted. Fairly frequent potting during early growth seems to benefit the young plants and encourages them to become sturdy and short-jointed. Finally transplant them into

13. *Sweet Pepper or Capsicum: a prolific cropper, ideal for cultivation in greenhouses, frames or a sunny border.*

6- or 7-in. pots before they become starved, and harden off gradually if to be grown outdoors. When in the greenhouse, keep the plants on the staging and not too far from the glass; syringe the foliage morning and afternoon and make sure the roots are kept moist.

For growing entirely under glass 7-in. pots are the most suitable and a rich compost is necessary. Frame protection is sufficient in the summer.

When planting outdoors, select either a warm border or a nice sheltered spot, and do not plant out until around the middle of June. A suitable planting distance is 15–18 in. from plant to plant.

For growing under cloches, the large barn type is necessary and the plants are best planted in a shallow trench and kept covered throughout their growth, raising the cloches up on the elevators as more headroom becomes necessary, some six or seven weeks after planting.

When planted out, the amount of watering should be gradually increased until the foliage practically meets in the rows, and then the whole area must be kept thoroughly moist until the plants are fully developed.

The fruits will keep for some time when they are ripe but naturally they are at their best when freshly gathered. The green peppers which have become so popular in this country during recent years are the red fruits picked before they completely turn colour.

The chief troubles of Capsicum are Botrytis, Green Fly, and Red Spider. In the case of the fungus disease, *Botrytis cinerea*, it is most important to practise strict hygiene and remove any diseased material which appears, at the earliest possible moment, and to keep the heat on and the vents open.

For Green Fly and Red Spider spray with Murphy Systemic Insecticide or give two applications of Murphy's Greenhouse Aerosol. Under normal conditions of growth, the first fruit should be picked about eight weeks after planting.

Two of the most popular and easily-grown varieties are Worldbeater and New Ace (F_1 Hybrid), both ideally suited for growing by the amateur for use either in a green fruited form or left to ripen to a bright scarlet colour.

Cardoon

Cynara cardunculus

PERENNIAL

The earliest mention of this vegetable was made by Parkinson in 1629 and it is said to have been introduced into this country about 1658. For many years past the Cardoon has been more popular in France than in England.

This subject is grown for its crisp succulent leafstalks which must be blanched in a similar manner to Celery. The stalks are stewed or may be used in soups or salads. In appearance the Cardoon resembles the Globe Artichoke and it has a somewhat stately look, with its grey foliage, when it is allowed to flower.

In a retentive soil, Cardoons should be grown on the flat, but the plant is somewhat thirsty and must be kept well watered.

On rather dry soils it is advisable to grow in trenches like celery, and the soil should be rich and well broken down if you are to obtain a satisfactory growth. Towards the end of April rows should be marked out 3 or 4 ft apart and groups of seeds sown at intervals of 18 in. in the rows. In due course, thin the plants to one per station and stake each plant when large enough.

Full growth is reached by August, when blanching is commenced by gathering the leaves together, wrapping them round with bands of hay, and earthing up. It requires eight to ten weeks to complete this operation.

The French method is quicker. Sow the seeds in pots under glass and in May the plants are put out 3 ft apart.

When fully grown, the plants can be 4 ft or more tall and should therefore be firmly secured to the stakes.

A covering of straw, 3 in. thick is thatched round each plant from bottom to top and each top is tied and turned over like a night cap. A little soil is then drawn to the foot, but earthing up is not necessary.

Carrots

Daucus carota

BIENNIAL

Few, if any, would question the popularity of this vegetable. The very fact that it is available for table use all the year round may have considerable bearing in this direction. It is easy to grow under cloches, in cold frames, and in cold houses; and various sowings can be made outdoors throughout the growing season, thus ensuring a continuous supply of young succulent roots.

Undoubtedly this subject has benefited from the vast vegetable breeding programme of recent years, and most varieties in cultivation today are practically coreless, as compared with some of the old varieties like James Intermediate and Long Red Surrey of bygone years. Further, the fashion in shape and size has changed considerably, the emphasis being placed on the finger-type Carrots which can be cooked and served entire, as opposed to the old method of growing the long varieties, removing the yellow cores, and cutting the root into pieces for the convenience of cooking.

SOILS AND MANURING   It is interesting to note that the best Carrots produced by commercial growers are obtained from light and sandy soils, such as will be seen in East Anglia, Holland, and other places. This does not mean, however, that useful produce cannot be obtained on heavier types of soil. Early digging, giving the winter frosts an opportunity to pulverize the land, and allowing the surface to dry out sufficiently before levelling off prior to drill drawing, will produce a perfect tilth.

This root crop should be grown on land which had been well manured the previous year for such subjects as Peas, Beans, Salads, etc. On no account should Carrots be sown on freshly-manured land. If, however, a doubt exists on the physical condition, a dusting of 2–3 oz per square yard, prior to seed sowing, of a complete fertilizer such as Growmore would be beneficial.

VARIETIES AND THEIR USES   A tremendous advantage can be obtained by using a suitable variety for a specific growing method as in the following examples. Where early Carrots are required, sowings can be made in the border of a greenhouse, in frames or Dutch lights, or under cloches any time from early February until April. For this method of cultivation, choose a variety such as Amstel, Amsterdam Forcing, or Champion Scarlet Horn, the first two named varieties being quick maturing with very fine foliage and a root about the shape and size of the middle finger, whilst the third variety, although the same shape, has slightly larger foliage and root, and is slightly later in maturity.

In the greenhouse or in a frame, seed may either be sown in drills 6–8 in. apart and thinned to the required distance, or alternatively broadcast thinly and raked in.

The approximate date produce is available for the table depends on sowing dates and the method of cultivation, but, to give a guide, if sown in February in a greenhouse border, roots should be ready for pulling in May; but, if sown under frames, Dutch lights, or cloches, to take the country as a whole, it would be June, this being a period of the year when a dish of young succulent carrots cooked and served entire enables one to appreciate fully how delicious this vegetable can be. Visitors to the continent will readily understand its value when cooked by the continental method with Peas and small Cocktail Onions.

In the past it was common practice to sow outdoors the whole year's requirements in one operation, the general idea being to use the thinnings in June and continue from the same bed, pulling as required, lifting the remainder of the crop, cleaning off and storing for winter use. It is not difficult to understand that in more northerly parts of the country, with a shorter growing season and colder conditions, this was the correct method, but with the newer quick-growing types, particularly, say, from Yorkshire south, the time has come when the method of cultivation should be changed. This serves two distinct purposes; firstly, it avoids occupying the ground for a whole season, secondly, it ensures young, first-class table material the

14. *Carrot: Champion Scarlet Horn. First early-finger shaped variety, almost coreless roots, excellent for successional sowings in frames or open ground.*

whole year round. The obvious answer therefore is to sow little and often. Where the grower is dependent entirely on outdoor growing his first sowing when soil and climatic conditions permit should be made in March, which will make table quality available in June/July.

Another small sowing should be made at the end of May, which would form a succession to the first sowing, covering July to September. A third sowing should be made in the north in early July, but in the south in the third week, to be ready from October and throughout the winter. Here again, it must be emphasized that early-maturing varieties are essential, such as Early Nantes and Champion Scarlet Horn, but, where a slightly larger root is required, Sutton's Favourite. Long types will be dealt with under exhibition notes, for the simple reason that this is generally the purpose for which they are required.

OUTDOOR SOWINGS   Distance between rows 12 in. is adequate, in fact commercial growers who grow for processing use a much closer row width and greater plant population in the rows. If the land is known to carry indigenous weed seeds, then the amateur grower would be well advised to sow at 1 ft from row to row to enable ease of hoeing and cleaning. The market grower has the advantage of chemical spraying, which is not available to the amateur for weed destruction; hence the denser plant population.

One of the troubles often experienced by the amateur grower is the Carrot Root Fly. When the plants are attacked by this, the first indication is the yellowing of the foliage, which goes on to take a purple hue. Unfortunately, when this occurs it is too late to do anything and often results in a failure of crop. It is timely, therefore, to suggest that, before sowing, the grower should purchase a packet of combined seed-dressing, and mix a small quantity of this with the seed in the packet, in other words, dress the seed; and, to make doubly sure after opening drill, dust with Gamma BHC Dust as a preventive measure. Seldom, if ever, does the commercial grower suffer from this trouble as he is wise enough to have all his seed dressed before sowing. Apart from this trouble, in general terms, this subject is seldom affected by others except root-splitting. This is not caused by pests or diseases and is generally found where a crop has experienced drought during the course of growth followed by rain, on land where a nitrogenous fertilizer has been used to excess. A well-balanced fertilizer, which includes potash and superphosphates, is more suitable for a Carrot crop; this should be worked into the soil before seed sowing at 2–3 oz per square yard. The potash will assist the plant under very dry conditions and will go a long way in preventing splitting of the roots.

EXHIBITION   To enable the exhibitor to produce first-class show material, cultivation methods are completely different, including preparation of the soil, soil compost, spacings, etc. A careful watch should be kept on the show schedule so that the correct variety is grown to meet the schedule requirements. Experience shows that, through a badly worded schedule or through the judges' lack of knowledge of varieties, unnecessary disqualifications of an exhibit have occurred. It is, therefore, important at this juncture to define the correct classification. The reason for making this point is that there is still a number of people in horticulture who think there are three distinct classes of Carrots, short, intermediate, and long; but such is not the case. The word 'intermediate' was attached to the variety many years ago, such as James' Intermediate, Scarlet Intermediate, New Red Intermediate, etc., but exhibitors, and in some cases judges, find it very difficult to interpret the schedule. A few years ago the Royal Horticultural Society in their wisdom published an article in their *Journal* for the guidance of all concerned, calling attention to the fact that there were only two classes of Carrot: a long-pointed variety, which embraces such varieties as New Red Intermediate, James' Intermediate, Scarlet Intermediate, Long Red Surrey, and Altrincham; and the remainder, commonly known as a short type, classified as 'other than a long-pointed cultivar', which are shorter-growing roots with blunt ends.

This subject forms one of the most important and most interesting at any show, whether in the single-dish classes or in collections. Its popularity, quite apart from its food value, is due to the high points value which it carries—which incidentally is fully justified, as a first-class exhibitor of Carrots requires more skill in cultivation than is generally appreciated. Experienced showmen have long since learned that to produce the ideal dish it is necessary to grow specially for exhibition purposes, and, to accomplish this, particularly for the long types, the ground is deeply trenched during the winter months, avoiding manuring. In early April the ground is raked down and levelled off, and, by using a crowbar, holes are bored to a depth of $2\frac{1}{2}$–3 ft, which are filled with a good, old potting soil. If this is not available, use a good sandy soil, mixing in plenty of wood ashes. This mixing should be undertaken in the winter months, and the mix kept under cover until required so that it is in a dry condition ready for use. When filling the holes before sowing, the soil should be pressed down by using a stake. For the long varieties of Carrots the holes should be 15 in. apart, but for the short-growing varieties 12 in. is sufficient. Sow two or three seeds to each hole—but remember to dress the seed with a combined seed dressing against Carrot Root Fly—eventually thinning out the seedlings to one at each hole when in the rough leaf state. During showery weather, a light application of a complete fertilizer will assist growth. Keep the bed clean by the constant use of the hoe, but at all times keep the crown of a plant well covered with plenty of soil to prevent greening which is a bad fault on the show-bench.

In the event of extremely dry conditions during the course of growth, it is advisable to keep the ground well watered as rains following the drought will cause splitting of the roots.

If growing the long types of Carrots, it is advisable to remove the soil at least a spit deep before attempting to remove the Carrots from the soil, as the hallmark of a well-grown long variety of Carrot is the unbroken tap root.

When lifting the short types of Carrots, ease the soil with the fork. After lifting, these should be carefully washed in clean water, using a sponge or soft cloth—on no account a brush, as this will scratch the skin of the root. After washing, lay out side by side in good light and select only the roots with the following features: uniformity of size and shape; absence of pale skin colour; absence of greening round the crown. Wrap each root individually in soft paper, and pack in a suitable-sized container for conveyance to the show.

Staging When staging, place lengthwise on the bench with the crowns furthest from you. Assuming the schedule asks for six roots, place three at the base, two on the top of this, and finish off with the last root on the top so that you have the formation of three, two and one. To stop the Carrots rolling on the show-bench, always take with you a little Carrot foliage and stalk and slide this under both sides of the bottom row of the heap. This method of staging can be applied to both the single-dish class and collections. Points value, 20.

47

Cauliflower

Brassica oleracea botrytis
cauliflora

BIENNIAL

It would be most difficult to find a more controversial yet more interesting subject. Its complicated sources of origin, times of maturity when grown in this country for both table and exhibition purposes, and a lack of complete knowledge of the subject are often the causes of disappointing results. At the same time, it is a very important vegetable, and the simplest manner in which we can best describe the uses of its many varieties is to divide them into groups. These groups will give the history, in addition to their uses when grown in this country, and will simplify the choice of variety for its specific purposes or time of maturity.

Many parts of the world played their part in the development, and, in so doing, suited the climatic conditions of the country of origin. By adapting the various groups, not only to various methods of cultivation but also to sowing dates, they have been dovetailed into a succession of crops covering most of the year. Provided climatic conditions are favourable, by using the various groups, we can ensure that there are very few weeks in the year, when we are without edible produce. Here it is necessary to state that geographical position, coupled with climatic conditions in any particular area, is the deciding factor: for example, the weather conditions in the northern half of the country would produce a longer gap in produce during the winter months than in the more southerly districts.

GROUP 1 Developed in northern Europe and Scandinavia, these are the dwarfest growing and quickest maturing varieties. Best results are obtained when seeds are sown under glass at the end of September or early October, over-wintering under glass, transplanting to the open ground, after hardening, in April, for heading in June and July. Alternatively, sowings may be made at the end of January under glass and transplanted to the open ground in April, for heading in June and July. This same group can also be sown *in situ*, in the southern half of the country, in early July outdoors, and singled down (not transplanted), thus providing a catch crop of quick-grown heads in October.

Recommended varieties which if sown and transplanted at the same time will form a natural succession: Classic, Snowball, Dominant.

GROUP 2 Mid-European origin—mostly French. This group may be sown under glass at precisely the same times as Group 1. However, they form a natural succession in the heading period when Group 1 is completed. The varieties in this section are All the Year Round and Lecerf. As the name All the Year Round

15. *Cauliflower: Classic. One of the most popular early croppings forcing Cauliflowers suited to sowing under glass late September–October or during January–February for summer cutting.*

suggests, this variety can be sown in a seed-bed outdoors in April and transplanted in June for August heading. In addition it can also be sown in May in a seed-bed and transplanted in July for late August/September cutting.

Group 3 Of Mediterranean origin, particularly southern Italy. These are the Giant Blue-leaved types, which for many years were so popular in this country for autumn heading, and included Autumn Giant, Dwarf Monarch, Beacon, Superlative and Autumn Protecting. These varieties can only be sown in a seed-bed in early May, transplanting to final quarters when ready in late June/early July for heading from September until the weather closes down in November or December, according to the area. It is important to say that, if these types are sown in April, they are more prone under certain climatic conditions to become interleaved or produce bracted curds.

GROUP 4 The most recent improvements in autumn-cutting varieties are those bred in Australia. Quite distinct in habit and growth from the Mediterranean types, being dwarf growing, they almost sit on the ground and are certainly more compact in habit of growth, and are much less prone to bracted or interleaved curds. To the less experienced grower, in the early stages after transplanting, they give the impression of very slow development, but once established they soon make the necessary growth to provide the perfect white curd, which is completely solid. Further, they leave little to be desired in the uniformity of the cutting period. It is, however, important to remember that, with the shorter cutting period, due to uniformity of strain, if Cauliflowers are required from September to the end of the season (i.e. November/December), the following range would be adequate, not only to keep the family fully supplied with fresh table material, but also to afford a surplus which can with confidence be used for freezing. A selection of

16. *Cauliflower: Beacon. One of the most reliable autumn heading Cauliflowers for September use. Sowing in a seed bed is usually best made during late May.*

the following varieties should provide an outstanding succession: Kangaroo for September/October; South Pacific for October; Barrier Reef for October/November; and Brisbane for November. The heading periods quoted are approximate and may be subject to climatic conditions and the geographical position in which they are grown.

GROUP 5 This group covers what are known as the Roscoffs, which were bred originally in the 1920s and subsequently, from the well-known French strain. It is, however, most important to note that this group is suitable only for very favoured areas, such as Cornwall and Pembroke, the coastal parts of which enjoy the benefit during the winter months of the Gulf Stream; coastal areas in Hampshire and the Isle of Wight; Devon; and, indeed, some varieties in east Kent, particularly Thanet. There is a lesser degree of success in these areas than in Cornwall or Pembroke, owing to the severity of the winter. In all inland districts, whether south, east or west, you are advised not to attempt to grow these varieties. In the Cornish Peninsula, say, from Land's End to Camborne and across to Truro, and in the southerly coastal belt of Devon, conditions are seldom as severe as in the remainder of the country, with the result that the heading season of the many Roscoff varieties continues non-stop from, say, December to April, while in all other districts growers can seldom cut a head in January or February.

GROUP 6   Best known for about 70 years as the Peerless Strains and in some parts of the country equally well known as the Feltham Strains, these were originally bred in Anjou in Western France, but further developed and acclimatized to suit English conditions. Covering about six varieties, they have been almost exclusively

17. Winter Cauliflower: Thanet. An extremely hardy winter Cauliflower which produces pure white heads during April.

used in the southern half of England in the past, below a line from the Wash to South Wales. The best varieties in this group are Superb Early White for January/February heading; Westmarsh Early for February/March; Snow White for March/April; with Summer Snow bringing up the rearguard to complete the cutting season in May. This group can best be described as of average type of growth, which produces a well-protected curd that is pure white and solid, in fact it would be true to say that many of the heads are almost completely wrapped by young leaves which not only afford protection from early morning frosts, but also assist in retaining the colour of the curd in sunny weather.

GROUP 7   This could well be described as the grandfather of all the groups of Winter Cauliflowers, and the only name suitable would be 'Old English'. Alas! many of the varieties which form this group have now been superseded by other, more recent introductions, but some of the best are still with us. In the colder and more exposed districts of the country, such as Lincolnshire, Yorkshire, and, indeed, further north, they still retain their popularity, being so adaptable to severe winter conditions, giving a very useful white head, and, as one would expect, later than other Groups in maturity. For example, Reading Giant, Continuity, Progress, Late Queen, and June Market are still in demand. Whilst these old hardy varieties still serve a useful purpose, one must not expect at all times quite the quality and colour of the new, hardy, headed strains referred to in Group 8.

GROUP 8 (Extra Hardy Types)   This group, of recent introduction, which is outstanding for inland, Midland, North Country, and especially exposed areas, is outstanding for depth, colour, and quality of curd. They have been proved in hard winters to withstand twenty degrees of frost, as against other varieties grown alongside which were severely damaged. Thanet is ready for cutting in April, White Cliffs April/May, and Manston, the latest to mature is ready the latter part of May.

In general cultivation the eight different groups to all intents and purposes cover the whole range; but for the benefit of future generations of growers there is another Winter Cauliflower which cannot be listed under the Sprouting Broccolis, and is quite distinct in its habit; it is therefore important to list it as Group 9.

GROUP 9 (Multi-headed Cauliflowers) In 1890 Suttons offered for the first time one such variety under the name of Sutton's Bouquet, but in recent years it has been known as Curtis' Nine Star Perennial. This has a typical Broccoli leaf, but the plant divides at a given stage of growth into several side shoots, which eventually develop a small white head on each shoot, in other words, several small heads on the one plant. After cutting the heads, dead foliage is removed, the ground is pricked over, at the same time being given a dressing of a complete fertilizer such as Growmore, when the plant will make fresh growth throughout the normal growing season, and produce a further crop off the old plants in subsequent years; hence the name, Perennial Broccoli.

SOILS AND THEIR PREPARATION The land on which SUMMER CUTTING VARIETIES (quick-maturing varieties) are to be grown (see Groups 1 and 2), must be in first-class condition, which means where fertility has been built up over a number of years and has received a coat of manure prior to planting. This is so necessary because on hungry land disappointing results may be experienced by poor development of the plant, which results in small immature heads. At the time of planting, use a balanced fertilizer, made up by weight with the following ingredients: 1 part Sulphate of Ammonia, 3 parts Superphosphates, and 1 part Sulphate of Potash, which should be applied at the rate of 2–3 oz per square yard. Transplanting distance should be 2 ft from row to row, with 20 in. from plant to plant. This, like Cabbage sown in January/February can become affected by Cabbage Root Fly. This is a timely opportunity for a reminder that prevention is better than cure; therefore, to ward off this trouble, dust the hole with Calomel Dust when transplanting.

Autumn Heading Types, represented by Groups 3 and 4, are also gross feeders; therefore, in preparing the land, a good coating of manure or compost will assist. Sowing for these two groups should be in a seed-bed in drills 1–2 in. deep, sowing thinly, we strongly recommend, to enable the young seedlings to develop into strong, healthy plants, with plenty of root growth. A thickly-sown seed-bed will only give under-developed plants which become quickly drawn and leggy, a condition which is detrimental after transplanting to good plant development; and experience shows that a Cauliflower which receives a check at any stage in growth is reflected in the quality of the curd. In bygone years horticulturists advocated sowing in April, but in recent years research has shown that the ideal sowing date in the seed-bed is anywhere in the first three weeks of May; but, if the weather conditions in more northerly districts are suitable, sowings in April/May would be beneficial due to the slightly shorter growing season in those areas. The one insect pest which can be troublesome on this subject as soon as the seedlings break the surface of the soil is the Flea Beetle. Here again, take preventive measures by dusting the bed with Sevin Dust as soon as the ground cracks. Oftentimes, the amateur grower may have blind plants, which only show after transplanting, due to the ravages of the Flea Beetle in the seedling stage.

To boost up these crops, the same dressing of a balanced fertilizer, as recommended for Groups 1 and 2, can be used to advantage, or a well-balanced propriety brand.

The Winter Cauliflowers in Groups 5, 6, 7, 8, and 9 should be dealt with in quite a different

51

manner to those already mentioned as regards the preparation of the soil. For example, the ground should be prepared, say, for early Potatoes, early Peas, Broad Beans, or Salad crops, which will have matured and been cleared before it is necessary to transplant the Winter Cauliflowers; manuring is, therefore, a very important item as the land will virtually carry two crops. Any residue, that is, unused food value not taken up by the first crop, will assist in growing the Cauliflowers, provided it is supplemented prior to transplanting by a well-balanced fertilizer. The seed is sown in a V-shaped drill, again thinly, to allow each plant individual root development. Having cleared from the ground the items mentioned previously, do not dig the ground, but create a tilth by hoeing and transplant. A useful hint here, and one well worth remembering, is that you will never get a reasonably-sized head or curd without the necessary stem development. Therefore, if you are faced with the choice of whether to put out early- or late-maturing varieties, always transplant the early-maturing types first to give them a longer season of growth before the weather closes down; whereas late winter and spring maturing types are far better planted later, having made about three quarters normal growth, so that, in the early months of the year when conditions are favourable, a top dressing with nitrogen and the use of the hoe will create spring growth and a first-class head.

With all autumn, winter, and spring heading Cauliflowers as represented by Groups 3 to 8, the final planting distances should be 2 ft between the rows, with 20 in. from plant to plant.

EXHIBITION   The experienced grower would be the first to admit that timing produce for a particular show is very difficult, due to varying seasons from year to year, together with the choice of a suitable variety for the desired purpose. It is, therefore, essential to pay great attention to growing a variety which will mature at the approximate date of the show. Often this is easier said than accomplished. The keen exhibitor will normally either grow two distinct varieties which head in natural sequence or alternatively he will use the one variety but stagger his sowing and transplanting dates.

The experienced exhibitor will, say, seven to ten days before the show, when the plants are producing perfect white, solid curds, pull up the complete plant, shake the soil from the roots, cover the head with tissue paper, and hang the complete plant up by the root in a cool cellar or similar place; it will remain in good condition until show day. If, say, two or three days before the show, the curds are in fine condition, break one or two of the outer leaves across the curd to exclude sunlight, thus retaining colour of curd, which is one of the important factors in a first-class Cauliflower. If staging in a collection on a backboard, uniformity in size of curd is of the utmost importance. Avoid any heads which show the slightest sign of bracting or woolliness, sometimes referred to as ricey. When trimming the curds for the backboard of a collection, trim the outside foliage back to below the level of the exposed part of the curd when staged, bedding the whole of the Cauliflower exhibit in Parsley. If showing in the single dish classes, be careful to read the schedule and interpret carefully what it asks for: for example, 'two Cauliflowers as grown', or 'heads of Cauliflowers', in which case the foliage should be trimmed back to about the level of the top of the curd. Points value, 20.

INSECT PESTS AND DISEASES   Insect pests which can affect the crop are mostly Aphids and White Fly. For both of these, spray with Liquid Malathion; this will, by killing off the Aphids, or preventing an attack, ward off possible virus infection, as it is the Aphids which carry the virus. For Flea Beetle, the treatment is as for other Brassicas.

Another common trouble found particularly in southern parts of England is Ring Spot. This can be to a large extent avoided, as it is usually found where too close planting has been practised, preventing a free circulation of air.

Club Root is another possible cause of disappointment, but this has been dealt with in another chapter (see p. 35).

It is important to ensure that the land where this crop is to be grown has received an adequate amount of lime, as Whiptail can occur where the soil has a high acidity. Whiptail results in malformation of the leaves, which become whip-like or strap-like, with much reduced and irregular blades of the leaves. The growing points become stunted and very often are blind. The curd is usually, small with leaves growing through it.

Celeriac or Turnip-rooted Celery

Apium graveolens Rapaceum

BIENNIAL

Celeriac is rather more easily grown than most types of Celery, as it is grown on the flat, not in trenches, and it produces a 'knob' or rounded root, weighing one or two pounds under average cultural conditions, and, of course, it stores more readily than sticks of Celery in the winter months. The root of the plant only is used and not the stalks.

Seed should be sown in gentle heat in a greenhouse in March. Afterwards, prick off and treat in a similar way to Celery; but after this the treatment is quite different. The plot of ground for Celeriac needs to be light and in good heart, and heavy land will need to be lightened by the addition of several inches of sand over the surface.

The plants should be put out on the level, 12 in. apart each way. Plant as shallow as possible, first removing any lateral shoots that could divide the stem. As with Celery, keep well watered, keep the crop clean by hoeing, and from time to time draw the soil away from the plants, for the more they stand out the better. Try to avoid any check in growth, or the plants

18. *Celeriac or Turnip-Rooted Celery. Excellent when stored for winter; use raw in salads or cooked as a vegetable dish, also in soup.*

53

will not reach a proper size. Any lateral shoots and fibres must be removed to keep the roots intact.

Towards the end of the season, cover the bulbs with a light coat of fine soil, and in December a portion of the crop may be lifted and stored in sand after the leaves have been removed, except those in the centre, which must remain. The remainder of the crop is not likely to need protection unless hard weather is experienced, when a covering of soil should afford ample protection.

Celeriac is cooked in the same manner as Beet and requires about the same length of time. The roots are trimmed, washed, and put into boiling water without salt or any flavouring and kept boiling until quite tender. They may then be pared, sliced, and sieved with white sauce or left uncut to be sliced up for salads when cold.

Celeriac is a good standby in the event of a hard winter when vegetables are in short supply.

Celery

Apium graveolens

BIENNIAL

This subject is a challenge well worth while to any grower, as it requires cultural skill and attention.

SEED SOWING For table produce from November onwards, there is no hurry for sowing; in fact, the best time is during March or early April, either in a flower-pot or seed-tray, using a rich fine sowing compost. Sow thinly and cover lightly, press down firmly, and water well, maintaining a temperature of 60°F. As soon as the seedlings are large enough to handle, these should be pricked out into fairly deep trays, keeping the young seedlings as near the glass as possible. They should be kept fairly moist and, when conditions permit, given some air—at first with caution—which can be increased as natural temperatures rise.

Preparing the Trench The earlier this is prepared the better. The simplest and most straightforward method is to take out a trench 18 in. wide and 1 ft deep during the late winter or early spring, placing at the bottom a good layer of well-made farmyard or stable manure or compost, covering this with about 3 in. of soil. The soil removed in making the trench should be stacked equally and neatly on both sides to enable a quick crop, such as Lettuce, to be grown and cleared before the soil is required for earthing-up purposes.

Transplanting into the Trench Care should be taken that, before transplanting into the open ground, the plants in the trays are well hardened by standing them outdoors for a few days in a sheltered position, to avoid the plants becoming chilled, which can result in their running to seed. To make it more easy in earthing at a later date, a single row down the trench planted 9 in. apart is strongly recommended. When removing the plants from the seed-trays, do this carefully with a trowel so as not to disturb unduly the root action, thus avoiding a check in growth. In the event of dry weather, watering is essential.

Earthing This should not be commenced until full growth has been obtained; and to acquire the necessary blanch for making the sticks tender allow a period of six to seven weeks. This sort of earthing-up needs considerable care. If the operation is to be carried out by the one person, the best method is to gather up the stems carefully, giving a light tie with either raffia or fillis string, which will enable the operator to use a trowel or hand-fork to place the soil loosely round the stem, to enable the heart of the plant to expand. After the first earthing, allow about two weeks, when the second earthing can be carried out by using a spade. Chop the earth over and lay it in heaps on either side of the plant. Then gather the stems together with both hands, remove one hand and with it bring the earth to the plant halfway round the base, and then, changing hands, pack up the earth on the other side. As soon as the plants require a final earthing, complete the operation; but at all times it is important to remember to avoid soil falling into the heart of the plant.

VARIETIES SUITABLE FOR TRENCH CULTIVATION These come into three categories: White, such as Solid White, strongly recommended; Pink; and Red. The Whites and Pinks mature approximately at the same time, whilst the Red is usually grown where Celery is required to stand in good condition until well into the new year.

Golden Self-Blanching Fifty years ago this form of Celery was very little grown in this country; it was more popular on the continent and in America. Nevertheless, its popularity has increased rapidly in this country in recent years, particularly where Celery is appreciated from August to October. It is very quick-growing and early-maturing, but its general cultivation is entirely different from trench-grown Celery.

PLANT RAISING Plant raising is identical to trench varieties. The preparation of the soil in the open is not difficult. Dig one spit deep, incorporating short, well-made manure or the compost heap; but, when transplanting, it is important to remember not to place a single row across the garden, as the secret of self-blanched material is in planting on the flat in a square block with a distance of 9 in. square within

19. Celery: Solid White. The best trench grown variety, also excellent for show purposes.

the block. Provided adequate watering in the event of dry conditions is available and an occasional liquid feed, which this subject appreciates, is given the plants will develop and touch each other, thus creating a straight stem which is well blanched. It is a fairly easily-grown crop, but it does require rich soil and moist conditions or the results can be most disappointing.

American Green (Tendercrisp) This type is grown on the flat in exactly the same manner and it produces pale green stems, which are not blanched in any way, of excellent nutty flavour and very crisp in texture when well grown.

EXHIBITION The keen exhibitor has long since realized that, in a Vegetable Collection, Celery is a 'must'. It forms, like the Leek, a good balance to the back board, and of course has a high points value. The two most popular exhibition varieties are Solid White and Un-rivalled Pink, the latter being outstanding, as it has the desired length and thickness of stick when grown correctly. Raising the seedlings can be undertaken in precisely the same manner as raising for trench Celery, but some experienced growers will prick off into single pots to avoid any check in transplanting. Sowing dates can vary slightly according to the actual date of the show. For August/September shows late January/early February is advised, but for a later show date, late February or March is preferred.

Preparation of the Soil This is really important. Double digging and, where possible, trenching, which means three spits deep, can be advantageous, incorporating a good layer of well-made manure, particularly in the top spit. Once again we give a timely reminder, particularly where early sowings are made, of the importance of carefully hardening the plants before transplanting. At no time should the plants receive a check in growth; therefore, take care

in transplanting not to damage the roots. Plant each plant on the flat (not in trenches) allowing a minimum of 18 in. from plant to plant, placing a stake or stout bamboo cane adjacent to each plant. When the plants have developed and have made sufficient growth to make it worth while, papering from the base should commence. Cut brown paper into the desired width, which should be placed round the plant, not too tight, placing about three ties, top, middle, and bottom, to hold in position. The width of the paper should correspond with the lengths of the stalks, to enable the foliage to appear above the top of the paper. At this juncture, it is well to remember that, prior to papering, careful examination of the interior of the heart should be undertaken to make quite certain that no slugs or earth worms have found a resting place. Make quite certain that adequate water, which is so essential, is given and that liquid feeds are given from time to time. As the plants develop and increase in size, it is necessary to remove the old paper. Again, examine whilst the opportunity presents itself, to make sure that pests such as slugs are not doing any damage. The above-mentioned procedure must be followed, increasing the width of paper as and when required. Check from time to time that the ties anchoring the plant to the stake or bamboo are secured. When changing papers, remove any side shoots or lateral growth at the base of the plant.

The hallmark of a first-class exhibit of Celery is a clean and unmarked length of stem, well blanched with perfectly healthy foliage, and, most important, a solid, compact heart. Lift carefully immediately prior to the show date. Put a light tie at the top of the stick just under the foliage before lifting to hold the stick firmly together, lift complete with root, wash off any soil left on the rootlets, remove the blanching papers, and trim off the rootlets.

After the thorough preparation of the sticks for exhibition, immediately cover the stems to avoid unnecessary exposure to light and the consequent risk of losing the perfect blanched effect. Points value, 20.

INSECT PESTS   Avoid slug damage by using either Slug Pellets or Liquid Slugit whenever necessary.

Celery Leaf Minor   During the course of growth this can be seen by a portion of the leaf turning yellow. Immediately this is observed, pick the portion of leaf off; if this is held up to the light you will at once see a small maggot. A dusting with old soot on the foliage from time to time should prevent the laying of eggs by the Celery Fly.

DISEASES   The chief trouble is Celery Leaf Spot (*Septoria*), which can and must be avoided. Prevention is possible, but cure impossible; therefore from about early August spray the foliage with Bordeaux Mixture, which is cheap and effective. This applies to all Celery, whether trenched, self-blanching, or, indeed, growing for exhibition purposes.

Chicory

Cichorium intybus

PERENNIAL

Chicory can be found growing wild in the calcareous districts of England, and bears a bright blue flower in the summertime.

The cultivated form of Chicory which we grow for salads was developed in Belgium. This type of Chicory is known as the Witloof or Brussels, and it is the roots of this that are forced or blanched to supply winter and spring salads.

Seeds should be sown in May, in rows 1 ft apart, and subsequently thinned to 9 in. in the rows. The soil should be deeply cultivated and rich, but not recently manured, except at a depth of 1 ft. The roots may reach the size of a good Parsnip.

In the autumn lift the roots undamaged, and only as required. Cut the foliage just above the crown, and cut all roots at the bottom to a uniform length of 8 in. They can be started into growth at once, but it is imperative that this is done in complete darkness. Put the roots in a dark cellar or shed in which the temperature is above freezing. They should be closely packed in deep boxes, with light soil or leaf mould between. If the soil is fairly moist, watering should be unnecessary for a month and had better not be resorted to until the plants show some signs of flagging. The blanched Chicory heads are often referred to as Chicons.

Breaking off the Chicons is preferable to cutting them, avoiding damage to the crown; if this method is used a second crop will be obtained from the same roots, but, as can only be reasonably expected, these will be of a slightly smaller size than the original. It should be possible to maintain a supply for salads from October until the end of May.

Chives

Allium schoenoprasum

PERENNIAL

A native of Europe, this is a plant which is frequently used as a mild substitute for the Onion in both salads and soups.

It will grow freely in any garden soil. Propagation is usually effected by division of the roots, either in the spring or autumn. The clumps should be cut regularly in succession, whether required or not, with the object of maintaining a continuous growth of young and tender shoots. At intervals of three or four years it will be necessary to lift, divide, and replant the roots on fresh ground.

Chives can also be raised from seed sown in March or April in drills $\frac{1}{4}$ in. deep. Thin the seedlings to stand at 6 in. apart. Thinnings can be transplanted if well watered. This method is cheaper than growing clumps or tufts, but it does mean waiting for a year or two for the seedlings to grow into clumps from which the leaves may be taken.

Corn Salad or Lamb's Lettuce

Valerianella olitoria

ANNUAL

Corn Salad grows wild in various parts of Great Britain and on the continent, and it is eaten by grazing sheep; hence its name—Lamb's Lettuce. The wild type looks rather like a large Dandelion, but the cultivated form has larger and more succulent leaves. As a raw vegetable it is not particularly palatable except when dressed as a salad with oil and the usual condiments.

Corn Salad is a quick-growing plant and is useful for its early appearance in the spring. The best sowings are made in August and September.

Seed may, however, be sown at any time from February to October, but only those who like the plant should trouble to grow summer crops, as, when Lettuces are plentiful, Corn Salad is not required. Most soils will grow Corn Salad but the site should be dry and open. Sow in drills 6 in. apart and thin to 6 in. in the rows.

The crop is cut in the same way as Spinach; either by the removal of separate leaves or by cutting over in tufts.

Cress, American or Land

Barbarea praecox

BIENNIAL

This is a native of Europe and is a variety of the Common Winter Cress which is a common weed in damp soils. The leaves of this plant resemble those of Watercress but the plant always grows on dry land and it is usually considered as a component of a winter salad.

The culture of American Cress is quite simple, as most soils will produce a good crop. For preference choose a moist spot, sow the seeds in drills which are about 9 in. apart. This is best done in late August or early September, when the crop will be useful in the early winter, especially if some form of protection can be provided to prevent frost damage to the leaves later in the season. It is advisable to remove the growing points of the foliage to ensure a continuous supply. Constant watering is necessary to prevent toughness of leaf. For a crop of American Cress in the summer months, a sowing would be necessary in March and June.

Cucumber

Cucumis sativus

ANNUAL

The Cucumber is a favourite in most households and as a greenhouse or frame crop it is fairly easily managed if certain rules and practices are followed in its cultivation.

There are two or three important factors in the cultivation of Cucumbers:

1 Seed sowing, and the first few weeks of the plant's life.
2 The preparation of the beds or permanent quarters.
3 The conditions necessary for the growing and development of the plant.

The first thing should be to clean and sterilize thoroughly the house where the propagating is to be carried out. The glass should be washed down inside and out to ensure the maximum amount of light for the young plants. Propagating of Cucumbers is usually carried out in the very early spring, when days are often dull and murky, and every bit of sun and light is valuable and must be made use of.

Young Cucumber seedlings require a temperature of round about 70°F with a minimum of 65°F. Make quite sure this minimum can be maintained whatever the weather.

CULTIVATION   Soil should be brought into the house to warm up in preparation for sowing. Some growers prefer sterilized soil; others prefer to use good clean turf and stable manure which has been staked out for a month or two; as an alternative the John Innes seed compost can be used. The soil should be fine, light and porous. Fill the seed-trays (previously sterilized) and press very gently with a flat board. Space the seeds on the surface about 1 in. apart, gently press in with the finger, and sprinkle a little soil over them—just enough to cover the seeds. Place the trays in the warmest part of the house and thoroughly soak, using a fine rose. Then cover with glass to prevent evaporation, and shade with paper. About the third or fourth day, according to temperature, the young plants will be touching the glass, which should be removed. The trays should be placed on the staging, where the seedlings will get the maximum of sun and light.

Potting should commence in about seven or eight days. Pots, usually 60s, should be well scrubbed or sterilized. Soil from a heap as previously mentioned is ideal and should be passed through a half-inch sieve, taking care to rub through as much as possible of the fibrous loam and manure; this should give a good, light and porous mixture. The plants should be inserted as low as possible in the soil, which should almost touch the seed-leaves. If the soil is in the right condition, no water will be necessary for two or three days, beyond a slight overhead sprinkle with a fine rose. Two or three sprinklings during the day, if the weather is bright, will be advantageous provided the plants dry off before night.

About the third day, or earlier in some circumstances, a good watering will be necessary. The water should not be too cold, and the pots should be filled; this should ensure moistening of the whole of the soil. Watering should then be carried on as necessary, for the soil should never become really dry. An occasional sprinkle with a fine rose is very beneficial during bright sunny weather.

Every morning, the paths, walks, and where applicable, pipes of the house should be well sprinkled with water to maintain the atmospheric humidity so necessary to the successful cultivation of Cucumbers.

Preparation of permanent quarters of beds While the young plants are being cared for in the propagating-house, preparations for planting can be made. The house should be

thoroughly cleaned and sterilized, as previously advised.

Making the Beds This is very important as on it depends the success or failure of the crop. The beds should consist of turf and stable manure in about equal proportions. The manure should be fairly fresh and very strawy, as this helps to keep the bed open for a long period and provides good drainage and aeration. The turf should be good and fibrous, and should be sprinkled with lime and a slow-acting fertilizer such as Bone Meal or Hoof and Horn.

Conditioning the Beds The beds are made and the young plants will be ready for planting in a few days. Warm the house and give the beds a thorough soaking; this watering combined with heat acts on the manure and the beds are very soon nicely warmed up.

When planting, insert the plants well down in the bed, leaving a cup-shaped hollow round the plant to facilitate watering. A good soaking should be given to settle the soil round the roots. No further watering should be necessary for a day or two, but every morning the paths and beds should be thoroughly damped down. This helps the plants and tends to check Cucumber's worst enemy, Red Spider.

The plants will very soon begin to grow—and grow rapidly. They must be kept tied to the supporting wires which run along inside the roof of the house. When the plant reaches the top, the point should be pinched out. All side shoots (and these are the fruiting growths) should be carefully tied to the wires and stopped at two joints; further laterals should be stopped at one joint. This treatment continues throughout the season.

Meanwhile, three to four weeks after planting, the cup round the stem of the plant should be filled with soil; roots from the stem soon form and are of great assistance to the plant when being heavily cropped.

Watering must be done regularly and the

60

beds must never be allowed to become dry. Cucumbers require plenty of nitrogen and fortnightly feeding with Dried Blood, Fish Meal, or similar feeds is necessary. Occasional top dressings of soil and stable manure spread thinly over the surface of the bed will encourage surface rooting and be very beneficial.

Atmospheric humidity is essential and this can be assured by keeping paths, etc. thoroughly damped down; and during hot weather spraying the plants overhead is of great assistance.

There is little doubt that the best method of growing Cucumbers in the greenhouse is by the bed system as already described in detail, but, for some amateurs with a small greenhouse and a household which requires quite a small quantity of fruit during the salad season, the use of 10-in. pots or fairly deep boxes placed on the stage of the greenhouse can be most successful for the plants to be trained up to the roof.

In Frames A 'hot-bed' is almost essential for frame-growing of Cucumbers, for such supplementary heat is generally necessary. For a 6 ft × 4 ft single-light frame, two plants will suffice, and these should be grown on small mounds to prevent water collecting around the base of the stems. When four leaves have formed, the leading shoot should be pinched out and the two lateral growths allowed to develop. These should be 'stopped' after four leaves have been produced. It is necessary to peg out these growths so as to cover the bed in a regular manner and any superfluous shoots which appear must be removed. Each fruiting shoot must be stopped at two leaves beyond the fruit.

Under Cloches The same general principles should be observed as given for FRAME culture

20. *Cucumber: Telegraph. The best variety for the amateur grower, suitable for the greenhouse, cold frames and for exhibition.*

but the two main shoots must, of course, be trained in opposite directions. Copious watering alongside the cloches must be given to ensure a plentiful supply of moisture at all times, and liquid feeding should be similarly carried out when cropping begins.

All Female Hybrid Varieties These require a little different culture. Seed should be sown as for other varieties in boxes in a temperature of 70°F, and the seedlings will be ready for potting into 3-in. pots in seven to eight days. The plants should be put into their final quarters at an earlier stage than with normal varieties, as these Cucumbers are inclined to produce fruit very early and therefore tend to neglect plant growth if left in the pots for too long. It is most important to have plants with a strong root system and to stimulate the vegetative growth at this period. All side shoots should be removed in the first 6–8 leaf axils of the main stem to avoid premature fruit-setting and to encourage growth, and it is most important that this should be done as early as possible.

Lateral shoots should be trimmed in good time, and this should be confined in the early stages to pinching off just the top of the shoots, whereas with normal flowering varieties the laterals are removed when the flower-buds start to yellow. The quality of the fruit obtained from the laterals is much higher, and only in the case of very vigorous growth should a few fruits be left on the main stem in the top of the plant. If the growth becomes too strong, the night temperatures can be reduced and this should help to stimulate fruit-setting. Provided no other type of Cucumber is grown in the vicinity there is no problem of cross fertilization and the

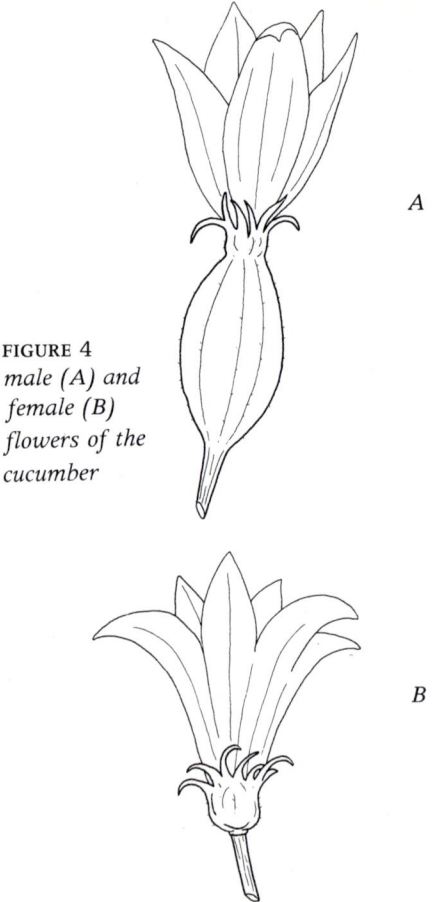

FIGURE 4
male (A) and female (B) flowers of the cucumber

production of bulbous fruits sometimes seen in normal varieties.

Male Flowers Occasionally a male flower may be produced, and any such flower specimens should be removed.

Cucumber, Ridge (Outdoor), and Gherkin

Cucumis sativus

ANNUAL

This group of varieties, which includes various distinct forms of outdoor Cucumbers and Gherkins, are all sun-loving plants and in reasonable English summers produce an abundance of good quality fruit in their respective sizes and shapes.

When there was an abundance of manure and it was easily obtained, many gardeners used the ridge and hill method of cultivation, removing the soil for about 3 ft wide and 2 ft deep and filling this area with well-made manure, making slopes from north to south. However, nowadays there is considerable difficulty in getting a supply of first-class manure, so many gardeners have been forced to use the compost which they have themselves made, adding to this a generous sprinkling of a complete well balanced fertilizer.

The time of sowing depends on the location of the garden. For the southern half of England, sowing outdoors should not take place until from mid-May onwards, and further north towards the end of the month would be early enough. Spacing can vary considerably, as in one season the growth may be rampant and plants will need to stand at 15–18 in.; in dull sunless seasons the plants are making little progress and could have been much closer; nevertheless, allow 6 ft for the trailing habit of the plant.

Many growers with glass protection sow in pots during the second half of April, for planting outdoors towards the end of May. For outdoor sowing, sow three or four seeds per station at 2 in. apart and 2 in. deep, and protect against slug attacks. When the young plants are in the third-leaf stage of growth, select the strongest pair of plants; the others can be either planted elsewhere or discarded. This operation should be in late May and, some weeks later in June, reduce the number of plants to one per station, which will allow of full plant development.

When the joint of the leading growth is 12 in. long, nip this out. By which time, the plant will have made six to eight leaves and this procedure will induce a branching habit. Do not stop any of the side shoots, though you can if you wish remove the weakest of them entirely. Should your plant continue to grow without producing a Cucumber, cut it back to seven leaves.

If you should notice fine white roots on the surface of the soil, cover them immediately with about a 2-in. layer of a good compost, press down firmly, and moisten at once.

Naturally, plenty of watering is necessary for the production of the best crop, but neither water nor fertilizer should be splashed against the stem, as this may cause rotting trouble at the neck of the plant and possible scorch of the foliage.

Besides watering the plants when necessary, it is essential to syringe both sides of the leaves to help create a moist atmosphere and also to discourage attacks by the Red Spider mite.

Should the leaves of your plants become yellow and puckered and the plants rapidly wilt, shrivel, and die, they have become infected with Mosaic Virus, which is carried by the Aphids. It is imperative, therefore, to keep them free from Green Fly by means of spraying with a systemic insecticide. If some leaves develop yellow blotches or greyish brown mould after damp muggy weather, cut them off and spray the plants with colloidal sulphur.

The Cucumber fruits should be cut whilst they are still a dense green colour and before they show any sign of yellowing at the stem end.

The use of cloches for Ridge Cucumber is of

immense value and, of course, it lengthens the season and, in many years, results in the production of much better quality fruit.

In recent years there has been a great deal of most useful work done by plant breeders in many countries, and we now have quite a wide choice of varieties, which produce heavy crops of almost greenhouse texture and quality; in addition, many of these are resistant to heat and powdery and downy mildew. Two of the best of these are Baton Vert and Burpless Tasty Green or Marion F_1 Hybrid. These are about 9 in. long, and crop well into the late summer. The variety, Perfection is unsurpassed, if one is content to grow this old and well-tried variety. One other outdoor Cucumber of rather unusual appearance is the Apple-shaped; this is extremely prolific, bearing fruits very little larger than a hen's egg and most attractive when used in a salad.

Gherkins are grown in exactly the same manner as the Ridge Cucumber and the fruits are used for pickling in a half-grown state. One of the best varieties in this group is Venlo Pickling which carries a tremendous crop of quite small, pale green fruits.

From time to time the question is asked: Why have my fruits a bitter taste? Generally the

21. *Cucumber: Apple Shaped. An unusual shaped Cucumber for outdoor sowing. Almost round, pale cream skinned; heavy cropper.*

reason can be traced to a sudden check during the growth of the plant. However, do not throw the fruit away: cut 1 in. off the blossom end, and rub this in a circular manner for about one minute on the cut portion of the remainder of the fruit. This will produce a white foam, which, when wiped off with a damp cloth, will remove the bitter flavour.

Dandelion

Taraxacum officinale

PERENNIAL

As a salad crop Dandelion has been used in this country for several hundred years; and French and Italian people have used green and blanched leaves of this as a delicious vegetable.

Most gravelly or chalky soils will grow good Dandelions and manure is hardly necessary if

the land is reasonably well dug. Sow seeds of the cultivated type during May and June and thin to 12 in. apart each way, and then keep the crop clean with frequent use of a Dutch hoe.

At any time during the winter, the roots may be lifted and forced in the same way as Sea Kale or they may be covered with pots in the spring to blanch where they are growing. In any case, the spring growth must be made in the dark for, if green, the flavour is bitter.

Those who like this as a salad plant may obtain early supplies by planting the roots in boxes in a cellar or other similar place and cover to exclude the light. Should the soil become dry, give a light watering; but not otherwise.

Egg Plant (Aubergine)

Solanum melongena ovigerum

ANNUAL

This is a native of South America and other tropical countries, but it has been grown in England since the sixteenth century, though very little for edible purposes. During the past few years there has been an increased interest in the Egg Plant in this country and now, throughout the year, fruits are on sale in many shops; and more gardens are growing a few plants in a small way, as happened with Green Peppers. There are in cultivation purple, black, and white fruited varieties, bearing fruits of various shapes, such as round and long.

From the point of view of cultivation, Egg Plants can be treated in a similar manner to Capsicums or Peppers, except that Egg Plants need little atmospheric moisture during seedling growth and the plants should be pinched out when 6 in. high. It is usual to limit the number of fruits to three or four per plant.

In most English summers, a cool greenhouse is most likely to be best for growing Egg Plants, although a warm, glass-sided frame is better than the majority of outdoor sites.

Sowing should take place in February or March in a heated greenhouse with a temperature of near 60°F for growing on in pots, or the border in a greenhouse or frame; but, if it is intended to grow them under cloches, delay sowing until mid-April so that the plants when hardened off can be planted out $1\frac{1}{2}$ ft apart early in June.

When the seedlings are about 2 in. tall, prick off singly into 3-in. pots, using J.I.3 compost, and, as soon as they begin to fill the pots with roots, move them into 6-in. pots. If they are to be fruited in pots, transfer to the $8\frac{1}{2}$-in. size.

When in full growth the Egg Plant will require plenty of water and frequent doses of liquid manure. Atmospheric moisture at this stage should be avoided, but, as soon as attacks of Red Spider are noticed, syringe the foliage with tepid water, including the undersides of the leaves. Also spray with a systemic insecticide from time to time, even if the plants are growing under cloches.

It will take about $4\frac{1}{2}$ months from time of sowing to the gathering of the first fruit. These should be removed as soon as they are fully coloured and while they have a polished gloss.

Endive

Cichorium endivia

ANNUAL

The Endive belongs to the same genus as Chicory and was cultivated from time immemorial by the Greeks and the Romans as a salad vegetable. In England it is known to have been cultivated in the sixteenth century. In this country we use the blanched 'hearts' as a salad, whereas in France they are also cooked and eaten as a vegetable.

SOIL   Heavy damp soils are unsuitable and by far the best is a medium to light, rich loam. Fresh manure is not recommended for the crop and it is advisable to use land which was manured for a previous crop.

SOWING AND TRANSPLANTING   For early crops seed may be sown in February in pots or boxes, and the seedlings pricked off and transplanted outdoors into their final quarters as soon as they are large enough, taking care to harden them off gradually. From these early sowings they can also be transplanted to Dutch lights, cold frames, or cloches, where they should mature quite well. By this method the plants should be ready for use in May; those outdoors will be a little later and will make a useful succession. For successional crops, sow outdoors from April onwards in rows 15 in. apart, and single down to 12 in. apart.

BLANCHING   Various devices may be used as it is the most important part in the successful development of this plant and its edible qualities. Cover the plants with large-flower pots with the drainage holes covered to prevent the light penetrating, or place old slates or boards over the plants, taking care not to use any material which has a tendency to stain the leaves. Plants must be dry prior to blanching.

As the blanched Endive plants will not keep long in perfect condition, because of a tendency to deteriorate and rot, it is advisable to blanch only a few at a time and in succession. In summer and early autumn blanching takes five to ten days, but later on it can take up to twenty days.

Slugs can be troublesome during the growing of Endive, and preventive measures should be taken early on.

The Curled Endives, such as Exquisite Curled are most suited to March–July sowing, and the plain-leaved types (Winter Lettuce-leaved) for sowing from July until late autumn.

22. *Endive: Exquisite Curled. The heads should be well blanched when full grown. A most excellent salad material when prepared and dressed in the Continental manner.*

Garlic

Allium sativum

PERENNIAL

Garlic was introduced into England from Sicily in 1548. It is not the same as our own Wild Garlic or Ramson (*Allium wisinum*) or Crow Garlic (*Allium vineale*).

Garlic does best in rich, light, sandy or gravelly soil and likes plenty of sunshine. The method of culture is much the same as for Shallots (*q.v.*) except that Garlic requires planting in February about 2 in. beneath the surface of the soil, and the bulbs may be closer together, about 8 or 9 in. apart each way.

When large bulbs are required for exhibition or other purposes, the cloves, as the divisions of each root are called, should be planted separately; but for general use, moderate-sized bulbs, planted whole, produce a heavier crop.

Gourd or Pumpkin

Cucurbita

ANNUAL

The Gourds, Pumpkins, or Squashes are natives of the warm climates and have been in cultivation for a considerable time. It is believed that they were first introduced to England about the middle of the sixteenth century.

By way of explanation, the term Squash often means the American counterpart of our Marrow and Squashes; Marrows and Pumpkins are all members of the Cucurbita family; but there is such a wide range of types and it is obvious that many of the varieties are not suited to an average English summer, as they tend to grow far too slowly in cool, sunless, damp weather conditions.

There are two distinct types of Squash cultivated in the United States—the winter and the summer. The winter ones include what we call the Pumpkins, and these are allowed to ripen on the plant, harvested before the early frosts, and stored under cover; while the summer squashes should be eaten young like our Marrows.

Under the most favourable growing conditions the summer Squashes (or Marrows) will crop in about fifty days; whereas the later varieties (or Pumpkins) will take up to ninety days.

CULTIVATION   The cultivation of Gourds or Pumpkins is precisely the same as for Vegetable Marrow, and, due to the uncertainty of the crop of most of the large-fruited, storing types, many English seed catalogues contain only one or two different types.

For the bush varieties a distance of 3 ft between plants is recommended, but with the trailers 5–6 ft is necessary. Stopping the plants is quite unnecessary with trailing varieties unless you are training them up posts, walls or trelliswork. Bush varieties of course never require stopping of any sort. An abundance of water is required throughout growth, and mulching is of the greatest assistance.

At the end of May in the south sow three seeds per station, pushing the seeds in edgeways and about an inch deep. It is preferable to sow *in situ* to avoid transplanting, as this family resents root disturbance. Some form of protection in the early stages is most helpful, covering either by cloches or even jam-jars. Keep a careful watch for slug attacks and use a slug destroyer early on. If roots appear on the surface cover with some good fertile soil. The large fruited Gourds or Pumpkins are excellent for storage purposes.

PESTS AND DISEASES The most serious troubles likely to be met with in general cultivation are attacks by aphids, which so often carry virus infection. On no account neglect to spray in the very earliest stages of growth, as so often little or no crop will result once virus commences to spread amongst the plants. At the first signs of mottled leaves, cut them off and burn without delay.

ORNAMENTAL FOR DECORATIVE PURPOSES ONLY One of the most attractive and decorative forms of Gourd is the Ornamental, which has a very fine range of small fruits of different shapes and colours, which ripen on the plants in the late autumn and can be used for indoor decoration in the winter months. These are not edible.

Herbs

The growing of most Herbs from seed is a straightforward and simple task. The plants come quite true from seed, and it is nearly always easier to raise your stock from seed than to propagate from slips or cuttings.

Whilst most gardens do not require a large plantation of Herbs, a moderate number is to be found in large and small gardens alike, and the tendency at the present time is towards an added interest in the kitchen, with the use of a wider range of Herbs than the usual plantation which contains only Mint, Parsley, and Sage.

The very popular Mint, Tarragon, and Lemon Thyme are not usually grown from seed. It is possible to grow French Mint from seed, but the result is by general consent not equal to the Mint grown from roots. However, other Herbs such as Basil, Borage, Chervil, Fennel, Marjoram, Parsley, Savory, etc., are easily grown from seed.

Both Angelica and Mint do best in a moist soil, but many of our Herbs, especially the aromatic ones, succeed in soil which is inclined to be dry, poor and somewhat sandy, rather than in soil which tends to be rich; but although in general highly fertile land is not necessary, sunshine they must have for their fragrant essences.

Allocate a narrow border with short rows not too far from the kitchen, and if possible sloping to the south near a path. Make a practice of early thinning; in most cases the thinnings can be transplanted if required. Keep the plot hoed and free of weeds and allow space for reasonable plant development.

Angelica
Angelica archangelica
BIENNIAL

This is a native biennial which is not easily raised from seed by an ordinary outdoor sowing, as the germination is frequently slow and irregular. Sometimes it will be several months before the seedlings begin to appear. The best results are obtained by placing the seed in sand, which should be kept moist for several weeks before sowing.

Uses The leaves and stalks are sometimes blanched and eaten in the same way as Celery and are also boiled with meat and fish. Occasionally, the tender stems and mid-rib are candied as a confection, while the seeds are used for flavouring liqueurs.

Balm
Melissa officinalis
PERENNIAL

Seed may be sown outdoors in May and the plants put out in the late autumn or spring months. It can also be propagated by cuttings.

Uses The foliage is widely used in flavouring soups and stews, sauces and dressings, and sometimes with salad; also in Balm Tea and Dandelion Stout and Wine.

Basil, Bush
Ocimum minimum
ANNUAL

This is a dwarf-growing aromatic Herb (6 in. in height), which should be sown in April in the open ground. It is mostly used for seasoning, and can be also be used for flavouring soups, stews, sauces, and salads.

Basil, Sweet
Ocimum basilicum
ANNUAL

A tender annual, originally obtained from India, this is one of the most popular flavouring Herbs. Sow in gentle heat February to March, prick out into boxes, and transfer to the open ground in June. May also be sown direct into the open ground in April and May. Thin the plants in the row to 8 in. apart, with the rows at least 1 ft apart. The flower stems must be cut as they rise and be tied in bundles for winter use. This practice will prolong the life of the plant until late in the season. It is used in the same way as Bush Basil.

Basil is generally used fresh but may also be dried after gathering in July and stored for winter use.

Bay Laurel
Laurus nobilis
ANNUAL

A hardy evergreen shrub or small tree with aromatic leaves. Propagate by cuttings of semi-mature shoots about 4 in. long inserted in a cold frame in August.

The leaves of this shrub are used for culinary purposes as a flavouring in soups, stews, sardines, figs, tomato juice, etc., and as an essential ingredient of a 'bouquet garni'.

Borage
Borago officinalis
ANNUAL

Height 12–18 in. Seed may be sown in the open ground late April to early May, and singled down to a distance of 15–18 in., with the rows 1½–2 ft apart as the plant is tall and strong in growth. Seed may also be sown in the autumn, and plants from this will flower in May; whereas those sown in the spring will not flower

69

until June. It is a hardy plant which thrives in poor, stony soil.

Uses The flowers are used for flavouring purposes, especially for Claret cup. It is also a great favourite with bee-keepers. The young leaves may be used in pickles and salads; and the leaves may be added to Pimms and fruit juices. The flowers dried are used for mixing as a pot-pourri.

Chervil, Curled

Anthriscus cerefolium

ANNUAL

Seed of this can be sown from April at intervals until the autumn. Height 16–18 in. In dry weather, frequent watering is necessary to prevent the plants from being spoiled by throwing up seed stems. For winter use, sow in boxes and keep in a warm temperature.

Uses The aromatic leaves are used for seasoning in mixed salads, omelettes, and soups. Excellent for garnishing in the same way as Parsley, but the leaves are more delicate and are better picked from the stem than chopped.

Chives

Allium schoenoprasum

PERENNIAL

This dwarf-growing subject is best propagated by the divisions of the tufts. These should be replanted every three to four years. It is also possible to raise Chives from seed sown in March or April.

Uses This Herb is related to the Onion and its shoots are used for flavouring soups, salads, etc., as *fines herbes* for omelettes and *soufflés*, and as garnish for all new vegetables. Essential

in *vichyssoise* sauce, sauce *tartare,* and with cream cheese.

Clary

Salvia sclarea

PERENNIAL

Seed of this should be sown in April, in drills 16–20 in. apart or in a seed-bed, transplanting the plants when large enough. Commence to harvest the leaves in August. After drying, rub up fine, and store in bottles for winter use. The fresh leaves may also be used for flavouring.

Coriander

Coriandrum sativum

ANNUAL

A feathery annual, with umbels of spicy seeds which are used in curries and casseroles, and ground over lamb and pork. It is sown in rows 1 ft apart in the early summer.

Dill

Peucedanum graveolens

ANNUAL

Seed may be sown in April in a shallow drill, 9 in. apart, and the seedlings thinned down to a final distance of 9 in.

Uses Can be used for flavouring soups, sauces, and pickles, and the seeds may be used as a condiment and in the pickling of Gherkins. Is used for making Dill Water, and for distilling to obtain Oil of Dill. For culinary purposes, gather the leaves as required. For other purposes, gather just before the seed is ripe. It has a delicate flavour and is used extensively in Sweden, Russia, and Germany with white fish, Potatoes, and Cucumber, and also in hot and cold dishes.

Fennel

Foeniculum vulgare

PERENNIAL

A hardy perennial which has been naturalized in some parts of this country. It is grown to furnish a supply of its elegant feathery foliage for garnishing and for use in fish sauces. Sometimes the stems are blanched and eaten in the same way as Celery and in the natural state they are boiled as a vegetable. The seeds are also used for flavouring. Sow in drills 15 in. apart in April and May and thin the plants to 12 in. apart.

Fennel, Florence or Finocchio

Foeniculum dulce

ANNUAL

A sweet-tasting herb which is very largely grown in southern Italy where it is eaten both in the natural state and also when boiled. If cooked in stock, it produces a sweet aniseed-like flavour and has a pleasant aroma. Fresh and unusual when used raw in salads. Sow outdoors in April in rows about 18 in. apart and thin or transplant to 6–9 in. When the base begins to swell, earth up the plants in a similar way to Celery. When transplanting, pinch off the tips of the root.

Horehound

Marrublum vulgare

PERENNIAL

Seed may be sown in April or May; thin the plants until they stand 15 in. apart. May also be propagated by cuttings.

Uses The dried leaves are used in making Horehound tea and Horehound ale; and it is also a well-known medical herb from which an extract is obtained for subduing coughs.

Hyssop

Hyssopus officinalis

PERENNIAL

May be grown from seed or plants. Seed should be sown in April; propagation is by division of the plants in spring and autumn, or by cuttings made in the spring and inserted in a shady situation. Plants raised from seeds or cuttings should, when large enough, be planted out 1 ft. apart each way and kept watered until established.

Uses This strongly-flavoured Herb is used in broths and stuffings, and the tender but bitter leaf and shoots are occasionally added to salads; mainly decorative.

Marjoram, Pot

Origanum onites

PERENNIAL

One of the most familiar Herbs in British gardens, its aromatic leaves are used both green and when dried for flavouring. Strictly, the plant is a perennial but it is readily grown as an annual. Sow in February or March in gentle heat, and in the open ground a month later. The plants require a space of 10–12 in. each way. Grows to a height of nearly 2 ft.

Marjoram, Sweet

Origanum majorana

ANNUAL

This should be sown in drills, 1 ft apart, during May. Thin the seedlings out to 9–12 in. apart. The flowers should be gathered dry and stored in paper bags, for flavouring soups and broths in the winter.

Lovage
Levisticum officinale Koch
PERENNIAL

Sow seeds in late summer, autumn or spring, like Parsley, it can be slow to germinate. It can also be propagated by the division of roots in the early spring. The plants should last for several years if the ground is kept well cultivated.

It is a tall perennial (up to 6 ft) resembling Parsley and Celery and liking deep, rich soil. The leaves and seeds are very pungent and taste like a cross between Celery and curry. Chop the leaves into salads, in meat and fish dishes; the seeds will give flavour to winter stocks and stews.

Mint
Mentha
PERENNIAL

Although four species of Mint have been grown in this country, the most popular for culinary purposes is *Spicata,* which is grown by division; planting takes place in February to March. Cover with about 2 in. of soil. It is advisable to replant every third year. Use in sauces and jellies, and for flavouring many of the vegetable dishes. The distilled oil is used for medicinal purposes, confectionery, etc., and also adds a fresh taste to salads in hot weather. Mint grows vigorously in damp soil but, if allowed to occupy the same plot of land year after year, the leaves become small and the stems wiry. However, the bed can derive much benefit from an annual top dressing of rich soil, applied towards the close of the autumn, when all remaining stems should be cut down to the ground.

Mint, French
Mentha Crispa
PERENNIAL

There are several seedsmen in this country offering a type of Mint from seed, as opposed to the one propagated by roots only, and this is known as French Mint. It is of a slightly paler green colour and the leaves are a little more puckered, but the flavouring compares favourably with the much more widely-grown type. Seed can be sown in the open ground in April–May.

Parsley
Petroselinum crispum
BIENNIAL

This is best sown annually from seed in March/April or May in rows 12–15 in. apart, and the plants singled down to 4–6 in. from plant to plant. If desired, the matured plants may be lifted in October and transplanted into frames to provide a supply during the winter months. Seed germinates very slowly—5 weeks or more before the seedlings are seen—and in dry weather it is most important to keep the soil moist. Frequently, the first sowing of the season is made in gentle heat in February and, when large enough, the seedlings pricked off into boxes for eventual transplanting to the open ground at the end of April, allowing each plant 12 in. apart each way. (See also p. 99)

Pennyroyal
Mentha puleglum
PERENNIAL

Best grown by obtaining plants from October to March. The plant has an agreeable odour, more powerful than Mint. The leaves, which are used for seasoning puddings and various dishes, should be gathered fresh as required.

Peppermint — *Mentha piperita*
PERENNIAL

This should be grown in the same manner as the common Mint. The leaves and stems are used for seasoning, but they are chiefly employed for the distillation of essences of peppermint.

Purslane — *Portulaca oleracea*
ANNUAL

Seed should be sown in drills 9 in. apart in April or May and thinned out to 6 in. from plant to plant. The young leaves and shoots may be eaten cooked as a vegetable, or may be used in a raw state as a salad.

Rosemary — *Rosmarinus officinalis*
PERENNIAL SHRUB

A hardy evergreen shrub which grows to a height of 5 ft. Easily grown from seed, the leaves are used for making Rosemary tea or relieving headache. An essential oil is also obtained by distillation. It should be grown in a sheltered place, as, although hardy, a severe frost could kill it. Best grown by planting in March or April. Leaves may be used for seasoning, but in the olden days the shoots were used for making hair washes, for flavouring lard, and for distilling to make oil of Rosemary. Good with white meats, chicken, and fish, and much better fresh than dried.

Rue — *Ruta graveolens*
PERENNIAL

A very decorative, hardy, evergreen shrub, very pretty in the Herb garden but not now used in cooking, the leaves having an exceptionally strong odour, which is disagreeable to most people. Sometimes used for seasoning and also for Rue Tea.

Sage — *Salvia officinalis*
PERENNIAL

May be grown from seed or plants. Seed should be sown in April and May, and plants should be put out at the same time. Gather the young shoots as required for seasoning and flavouring. To dry Sage for winter use, tie in small bundles and hang in an airy loft or shed. When dry, strip off the leaves, rub up finely, and store in airtight bottles. Use carefully as it can be overpowering. Good with goose, pork, or duck.

Savory, Summer — *Saturela hortensis*
ANNUAL

Seed should be sown in shallow drills in April and singled down to 6 in. apart. The leaves and young shoots are used for flavouring soups, meat dishes, stuffings, etc., and for other culinary purposes. Slightly bitter, similar to sage. Traditionally used in France with Broad Beans. Good with fish, cheese and egg dishes.

Savory, Winter — *Saturela montana*
PERENNIAL

This hardy dwarf evergreen is propagated by cuttings, but can also be grown from seed, sown in the late spring. Its uses are similar to Summer Savory.

Sorrel — *Rumex scutatus*
PERENNIAL

Seed should be sown in April in drills 1 ft apart.

Single down 3–4 in. from plant to plant. Sometimes cooked and served like Spinach, it is also used as an ingredient in soups, salads and omelettes, and as a garnish for fish and veal. Slightly acid flavour.

Tarragon
Artemisia dracunculus
PERENNIAL

Must be planted in March or April. It is used in salads and soups, and the tops are pickled with Gherkins. Also used for making Tarragon vinegar. A favourite herb in French cooking, it has an especial affinity with chicken; essential in *sauce bearnaise* and *sauce verte*.

Thyme, Common
Thymus vulgaris
PERENNIAL

Seed should be sown outdoors in April, or plants put out during the summer months. The aromatic leaves and young shoots are used for seasoning, and always in *bouquet garni*. Pungent

and good in recipes, including wine; with butter, to garnish vegetables; in fish chowders; also good with chicken and turkey.

Thyme, Lemon
Thymus citriodorus
PERENNIAL

Must be grown from plants, which are usually planted in the summer months. The leaves and shoots are used for seasoning purposes. Specially recommended for drying and bottling for winter use. It is also used for scenting soaps and may be used for filling sachets. Delicious with eggs—omelettes, scrambled eggs, fillings for egg sandwiches, etc.—and with grilled fish.

Wormwood
Artemisia absinthium
PERENNIAL

This medicinal perennial grows 3–5 ft high and is best grown from plants. Sometimes used for flavouring, but mostly used in the manufacture of various kinds of liqueurs.

Horseradish

Cochlearia armoracia

HARDY PERENNIAL

Horseradish is to be found in various parts of Europe and in this country is seen growing wild on light soil.

This subject requires an open sunny position and well-manured, medium to light soil for the best results. In the absence of farmyard manure, do not hesitate to use well-made compost in its place.

Horseradish roots (which are known as thongs) can often be purchased from a well-stocked garden centre; the best roots are usually 8–12 in. in length. A good method of planting is to bore holes and plant the roots vertically 1 ft apart each way; this can be carried out in late February or early March according to soil and weather conditions. By the following autumn these will become large succulent sticks, far superior in quality to the poor material usually grown under starved conditions. If you wish, the roots may be dug as required, but a much better practice is to clear the bed at once and store the produce in sand for use as required.

This plan of campaign should be repeated each year, if possible on a fresh piece of ground each year. This change of site is beneficial as Horseradish has a great tendency to spread over a wide area, often to the detriment of other crops. If, however, you wish to avoid the necessity of making a new plantation year by year, it is possible to enclose an area of ground with polythene sheeting which will prevent spreading of the roots. Obviously, after you have obtained your first stock of roots, each following year the new plantation can be planted with thongs from your own roots, similar to those you purchased.

When the roots are lifted, the tops are cut off and the roots stored under cover in slightly moist sand or soil. When required they can be taken out, scrubbed, and grated just before they are required for table use. It is possible to shred the roots, dry in a slow oven, and keep in a corked jar; but these are not equal to fresh Horseradish.

Kohl Rabi

Brassica oleracea caulorapa

BIENNIAL

Kohl rabi is a widely-grown vegetable on the continent and it is to be seen in the markets of most of their cities and large towns.

SOILS Any soil which will produce good Turnips should certainly give a worthwhile crop of Kohl rabi as its manurial requirements are much the same. As for most other root crops,

the soil should be in a well-balanced state of fertility to produce young, tender, quickly-grown roots.

TIME OF SOWING Seed may be sown from March to early August in rows about 15 in. apart. Thin the seedlings in the early stages of growth to 3 in. apart, and, as the plants develop,

23. *Kohl Rabi: Earliest White and Earliest Purple. A fast-growing root vegetable with a pleasing Cabbage/Turnip flavour, especially if eaten before full maturity.*

a further thinning should be made to leave them at a distance of 6 in. apart. When the roots are quite small but actually large enough to be useful in the kitchen, remove each alternate root for the remainder to mature fully at 9 in. apart.

Kohl rabi will grow quite as quickly as early varieties of Turnips. The roots are best used before they become large, coarse, and pithy. There are two distinct varieties grown in this country: the Earliest White and its purple counterpart, the Earliest Purple.

The plant is quite hardy, and late-sown Kohl rabi may be left to stand outside all the winter unless the weather is really severe. It does not, however, store well under cover for any length of time.

Because of its very rapid growth, this subject makes an excellent catch crop in the late summer period—sown say from mid-July to early August for consumption during November and December.

76

Leek

Allium porrum

BIENNIAL

The Leek has been cultivated from the very earliest times, but the date of introduction into England is believed to have been about 1562. Gerrard, who wrote in the same century, remarks that 'Leckes are very common everywhere in other countries as well as in England.' In addition to this, the Welsh people adopted it as their national emblem from even earlier times, when they used the wild Leek as a distinguishing badge.

The Leek is one of the most popular winter vegetables grown throughout the British Isles, and in recent years the season of use has been extended by the introduction of earlier varieties, so that it is now possible to commence lifting Leeks in the early autumn.

Most types of soil will suit Leeks, but probably the best is of a light sandy nature enriched with well-rotted stable manure, or, failing this, a well-made compost supplemented by a complete fertilizer.

TIMES OF SOWING   To obtain the largest and finest specimens of the highest quality, seed should be sown in January and February under glass, and this, of course, is imperative for the production of perfect specimen stems for exhibition, as the plants must have a longer season of growth than is generally allowed for ordinary crops for kitchen use.

Sow the seeds in moistened soil in a pan or pot in a greenhouse with a temperature of around 55°F. Cover the seeds very lightly with fine soil and, as soon as the seedlings reach about 2 in. in height, prick them out into shallow trays filled with a good rich compost, allowing them 3 in. apart each way, or, to make an even better plant for exhibition, prick off each seedling into a single pot. Great care should be taken not to break the one slender root on which the plant

depends at this stage of growth. After pricking out, grow on in the same temperature until about mid-March, when they will need to go out into a cold frame to commence the hardening-off process in readiness for final planting out when the soil and weather conditions are suitable in April.

24. Leek: Prizetaker. A fine, long-shafted early Leek, outstanding for exhibition and table use.

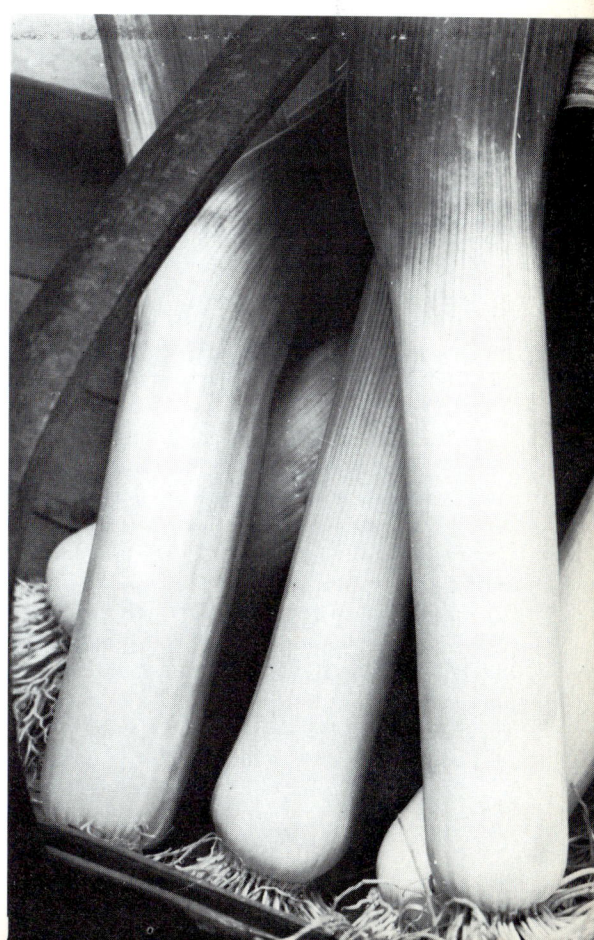

Most exhibitors grow on the flat and use a drainpipe or cardboard tube, not less than $2\frac{1}{2}$ in. diameter and 12–15 in. long over each plant as soon as it has made sufficient growth. Some exhibitors buy paper collars and tie them round the stem with raffia, adding a second collar when the first becomes outgrown; or you can move the first collar upwards and earth up the part already blanched.

When Leeks are required for general use in the kitchen with no thought of exhibiting in mind, seeds may be sown in the open ground during February, March, and April, but in most gardens the best time for sowing is about the middle of March. Generally speaking, transplanting takes place in June and July, often after one of the early summer crops, such as Peas, Broad Beans, or early Potatoes has been cleared. When the young plants are some 6 in. high, they are suitable for moving. Well-prepared land is required and a good soaking is imperative, as the move often takes place in the midst of very hot weather.

Average planting distances are 6–9 in. between the plants and $1\frac{1}{2}$ ft between the rows. Plant with a dibber and put the plant as deep as the base of the leaves.

Most gardeners take out a shallow trench about 1 ft deep and 1 ft wide, in which manure or compost has been dug, and, as the plant grows so the soil is brought up the stem to produce the blanched portion which we eat.

On the other hand, many growers prefer to make holes with a dibber about 2 in. across the top and 6 in. deep and drop one plant into each hole, filling each hole with water after planting. In this way the water will wash sufficient soil over the bottom of the plant to get it growing and the remainder of the hole will be filled in by the weather during the season. Perhaps it may be necessary to pull the plant up straight from time to time. If you adopt this method, the best distance between the holes is 9 in.

It is quite a good plan at planting time to trim the roots to about 1 in. or so of the base of the

plant and cut off the leaves square at about 4 in., so that the ends do not decay before recovery from the check.

Pot Leeks These are quite distinct from our usual idea of a Leek, as they are the short thick Leeks so widely grown by miners and others in Durham and Northumberland. They are not grown for show to produce a long, white, blanched stem, but are short and thick, with what is known as a 'tight button' at the V of the lowest leaf, which must be from 3 to 6 in. from the base of the stem. Sometimes a 6 in. stem can measure as much as 14 in. in circumference and contain over 100 cubic inches, but it is of perfect texture and quite free from coarseness, due to the manner in which it has been grown.

The choice of variety of Leeks grown under ordinary garden conditions and purely for edible purposes is simple and straightforward. For exhibition purpose, and for autumn and early winter table use, through October, November, and about mid-way through December, Prizetaker is outstanding. It is one of the slightly paler green-foliaged varieties, with a long, easily-blanched stem and a mild, agreeable flavour. To follow this in time of maturity, there is that well-established variety, Royal Favourite, which bears dark green foliage and is very hardy and longstanding—a fine Leek for spring use.

EXHIBITION Great care and patience is necessary to avoid even the slightest blemish appearing on the specimen stems. Firstly, gently remove any soil, leaf-mould, or peat attached to the stems. Drain pipes, if you have used them, must be carefully removed; but paper collars or cardboard tubes are usually left on until after lifting to help protect the stems and then cut down vertically so that they can be taken off without damage to the stem.

After lifting, wrap the plant in clean, damp paper, take under cover, and stand the roots in water until you prepare the stems for show. Be careful not to expose to the light or the stems will lose their whiteness and tend to go green. The final preparation will mean gentle washing and rubbing the roots to clean them thoroughly and then sponging of the leaves. Remove with the greatest possible care any loose outside skin if it is absolutely necessary, but keep this operation to the minimum.

Finally, tie round each stem at the base of the leaves a strand of wide raffia to prevent the leaves splitting down the stem due to the weight of the foliage. Points for judging, 20.

Lettuce

Lactuca sativa

ANNUAL

This popular salad plant was cultivated extensively by the Romans, but it was not until the reign of Elizabeth I (1562) that it was grown in this country. Records show that, thirty years after its introduction, the number of varieties had reached a total of eight. Comparison with the present-day list of varieties must surely indicate how important, and indeed popular, this subject has become.

However, it can be claimed that during this present century great strides have been made in producing many new varieties, each having special characteristics which dovetail into methods of cultivation, sowing times, variety of types of soil and district, consumers' preference for texture and flavour, resistance to diseases, such as mildew, virus, etc.

It can be safely assumed that at the time of introduction of this subject into this country the only crops grown were in the open ground, using successional sowings under cloches, frames, and in cold and heated houses. Varieties have been bred for hardiness to stand throughout the winter in the open ground; whilst in recent years so much research has resulted in considerably reducing the risk of crop failure through Mosaic Virus. On reflection, the grower of today has a first-class choice of variety for the desired purpose, which means that, provided glass is available in the winter months, there is no reason why the household should ever be without a supply all the year round.

To qualify this statement, it is necessary to divide this subject into various groups and methods.

Group 1   Successional sowings from March–August outdoors.

Group 2   Autumn sowing outdoors to overwinter.

Group 3   Sowing under glass in January–February to transplant to the open.

Group 4   Under glass August–January to mature in cold or heated house.

Group 5   Leaf Lettuce.

Group 6   Cos.

GROUP 1: Successional Sowings Outdoors Almost all types of soil are suitable provided a fine tilth can be obtained for opening up the drills, which should be as shallow as possible—say, about one inch deep. Select a portion of the garden which was well manured for a previous crop, the ideal being after such subjects as Peas, Beans, Celery, Leek, or Onions. If, however, you have any doubt concerning the fertility of the plot which you propose to use, broadcast 2–3 oz per square yard of a complete

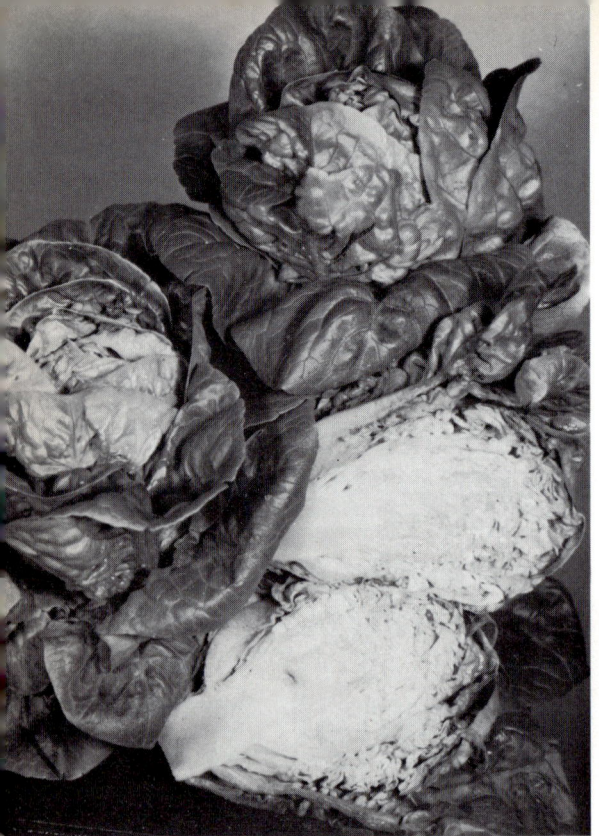

25. *Lettuce: Little Gem. The earliest and most compact of all Cos Lettuces, intermediate in character between a Cabbage and a Cos, extremely crisp texture.*

fertilizer such as Growmore and rake in before sowing. It is opportune to mention at this juncture that best results will be obtained by sowing *in situ* and singling out to the required distance. For the butter-head types (see illustration), allow 10 in. apart, and for the crisp heart varieties such as Webb's Wonderful, 12 in. apart. If, however, the small-growing Cos type is preferred, use Little Gem, as this variety is so compact in growth with very little outside leaf and it is only necessary to single to 4 in. apart. The one mistake that the average amateur continues to make is sowing too big an area at any one time, which only means that the

household are unable to use all the heads produced before they either run to seed or commence to rot. The answer to this problem is perfectly simple—sow little and often, say, every 3 weeks from March when soil conditions permit, until the first week of August. For example, to keep a family of four supplied with fresh produce a 5-yard run of drill would be ideal.

GROUP 2: Autumn Sowing. Over-Wintering Outdoors A sowing made of a hardy variety such as Valdor or Imperial Winter is well worthwhile for the simple reason it provides for the grower without glass the first heads of the season, which are always so much appreciated by the household. In this case, however, the seed is sown in a seed-bed in September, in northerly districts the early part of the month, and in the south the middle of the month is ideal, the object being not to get the plants too large or succulent for over-wintering. Indeed, if a few weeds appear, leave them in the bed for added protection during the winter. Lift plants carefully with a trowel, transplanting to their final quarters as soon as soil and weather conditions permit in the early spring. Distance between the rows 12 in., with 10 in. from plant to plant. This crop does not require land that has been heavily manured during the winter months; provided it is in reasonable condition, 2–3 oz per square yard of Growmore fertilizer worked into the soil prior to transplanting will under normal circumstances be sufficient to carry the crop.

GROUP 3: Sowing under Glass in January–February to Transplant in Open Ground Where a greenhouse is available, excellent heads can be obtained in May and early June by sowing seed thinly in a seed-tray filled with John Innes Seed Compost No. 1 and, as soon as the seedlings are large enough to handle, prick out into further seed-trays; a temperature of 50°–55°F is ideal, but, a week to ten days before transplanting to the open ground, the plants should be well hardened. The ideal variety for this method of cultivation is Fortune, which may be a day or two behind Premier, but it will produce a little larger head. Planting distances are the same as those given for the autumn-sown varieties in Group 2.

GROUP 4: Cultivation entirely under Glass for Winter and Early Spring Use In commercial circles these short-day, forcing varieties are an established fact and are recommended for the amateur grower where a border in the green-house is available. It is, however, most important to be careful in choice of variety. For example, for the cold greenhouse for an August or September sowing for November or December cutting, the variety Kwiek is outstanding, whilst the variety Kloek is more suitable for the cold or slightly-heated greenhouse for a mid-October sowing for March maturity; where heat can be maintained throughout the winter, use the variety Kordaat. The planting distances for both Kwiek and Kloek should be 10 in. × 8 in., but for Kordaat, which is slightly smaller when matured, 8 × 8 in. Normally, these short-day

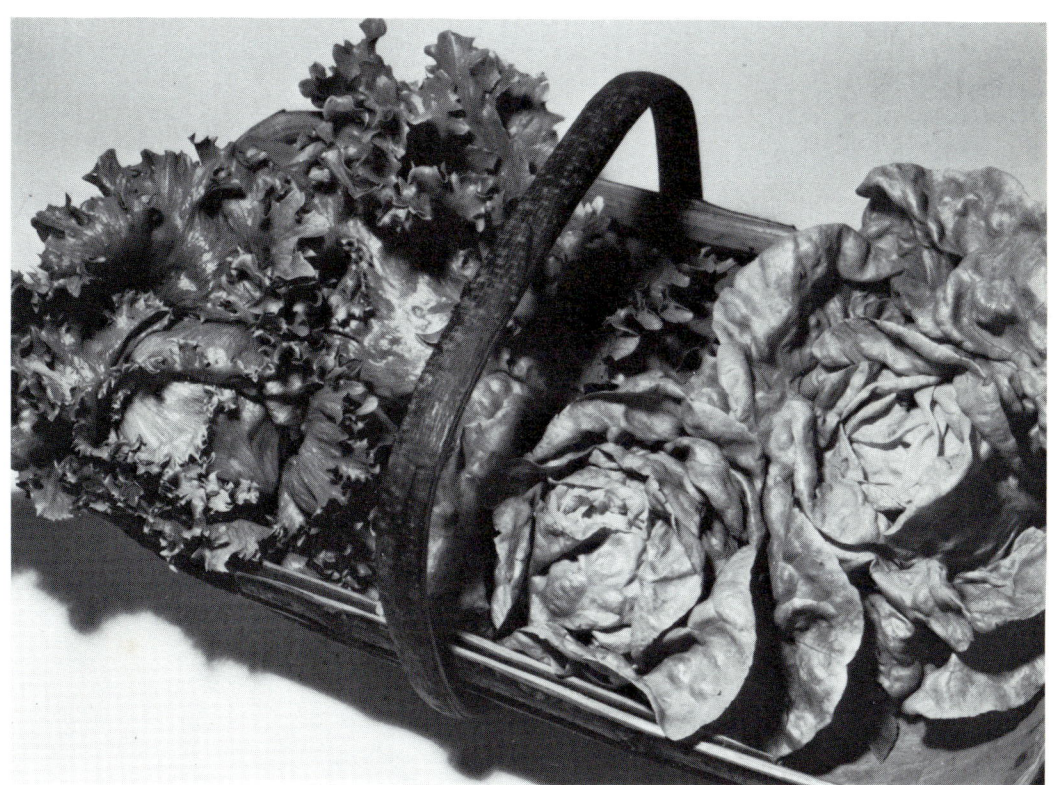

26. (Left) Lettuce: Windermere. The earliest maturing crinkled-leaved summer variety, very slow to bolt under hot dry conditions. (Right) Lettuce: Fortune. A quick-growing 'butterhead' type suited to early sowing under glass for transplanting outdoors or for spring and summer sowing in the open.

Lettuces follow a crop of Tomatoes or something similar, which means the soil is in good condition, probably carrying a reasonable residue of manurial value from a previous crop, which should be sufficient to carry the intended Lettuce crop.

A common failing of the amateur grower who is growing Lettuce under glass is over-watering, particularly bearing in mind the shortest hours of daylight. Two days before transplanting Lettuces, give the ground a good soaking. This should be sufficient to get the plants well established, and further watering should be as light as possible and only when really necessary.

Some considerable measure of disease control can be established by dusting the plants at weekly intervals after planting out until the plants touch each other, twice with Thiram and further alternating once with Zineb and once with Thiram.

In the cold greenhouse, if a severe frost is threatened, do not hesitate to lay sheets of plastic on the plants, removing it during the day and replacing it in the evening. Another timely reminder: ventilation of the house is most important, and this will go a long way to preventing Mildew.

GROUP 5: Leaf Lettuce A typical example of this type of Lettuce is Salad Bowl which has been offered for the past few years and has become quite popular, as the leaves can be gathered as and when required; use in salads or for garnishing, leaving the plants to produce a continuous supply. In point of fact this type of Lettuce is not entirely new. As far back as 1909, Suttons offered Winter Gathering, a slightly more curled form of the present-day Salad Bowl; and in America and, indeed, parts of the continent quite a number of these varieties are still in being. It is interesting to note that years ago they were referred to as Chicken Lettuce. These are best grown by sowing in situ in rows

1 ft apart, singling down to 12 in. from plant to plant.

GROUP 6: Cos Lettuce Without going into detail, the improvement in varieties in recent years has been outstanding in one important direction, that is, the evolution of a good self-folding type. Fifty years ago it was a common practice for a grower to tie the plant round with a strand of raffia to form a compact heart. These varieties are no longer with us; a well-grown crop is completely self-folding, particularly a variety like Lobjoits Green, provided checks in growth in the early stages are avoided. One of the chief causes of checking is failure, when grown in situ, to single the plants in the early stages of growth; another cause is attempting to transplant in the middle of the summer when conditions are hot and dry.

For a specimen of the self-folding type to be

27. Cos Lettuce: Lobjoits Green. The perfect self-folding type, crisp leaves, and resistant to bolting under hot dry conditions.

82

well grown it must have adequate spacing. To obtain this, allow 15 in. between the rows and 10–12 in. between plants. For the small garden, where supplies are appreciated throughout the growing season, it is more profitable to grow the small type like Little Gem for successional sowing from March to early August, and Winter Density, which is darker green in colour and perhaps a shade larger when fully matured, for a September sowing for over-wintering, transplanting in February/March for May cutting.

EXHIBITION   All show schedules have a class for this subject, but it would be very difficult for anyone to lay down hard and fast rules for sowing dates, as this subject depends on many factors, such as variety being grown, varying climatic conditions, etc. The best advice, therefore, that can be offered to ensure produce in prime condition on show date is to make successional sowings which, if one particular sowing has passed its best on show day, the next sowing should be in prime condition; to meet this requirement, 14-day intervals between sowings should meet the case. When selecting and staging specimens, condition is all important, so avoid a fully-matured head which is going slightly pale, as judges in examining, particularly in humid weather, will often find sliming in the heart. So have yours slightly under matured rather than over. For transport-

ing the heads to show, most exhibitors will tie damp moss or damp newspaper around the roots to help preserve the freshness of the Lettuce until the specimens are staged. Points value, 15.

DISEASES AND INSECT PESTS   Crops grown outdoors are reasonably free from pests with the exception of Green Aphids, which are not difficult to control, but in drought conditions in some seasons Lettuce Root Aphids can be troublesome. When these are experienced, it is advisable to pull up the plants and burn them. Another reason for carefully watching for Aphids is the prevention of the spread of the Lettuce Mosaic Virus. In the past few years, seedmen have spent a great deal of time and money in producing clean seed. Nevertheless, in spite of these efforts, many bad attacks of virus can be traced back to the lack of precautionary measures on the part of the grower, particularly where successional sowings are made and debris left on the ground, including virus-infected plants. In your own interests, all debris should be removed and burnt to prevent the spread of the virus, which is carried out by Aphids.

Another trouble which can be experienced, but, thank goodness, not very often, is Tip Burn, in which the edges of the leaves on a semi-matured plant take on a brown appearance, brought about by scalding sunshine after a shower of rain.

Marrow

Cucurbita pepo ovifero

HALF-HARDY ANNUAL

There seems to be considerable doubt as to when or how the Vegetable Marrow was introduced to the British Isles, as, prior to 1816, the name 'Vegetable Marrow' referred to the

fruit of the Avocado Pear. It is possible that some of the squashes of the U.S.A. found their way into this country, and that either planned breeding or accidental cross-pollination took place to produce the varieties which are now typically English and are seldom used in other countries.

At any rate, the Marrow has established itself as a popular English summer vegetable which is produced without too much effort from July to the end of September. In recent years there has been a tendency to grow the smaller-sized varieties, except perhaps for the show exhibitor who is trying to grow 'the heaviest Marrow in the show'.

The smaller types, such as the Improved Green Bush, are extremely prolific in bearing, at least three weeks earlier than the trailing types, and, of course, require much less space for their cultivation.

SOILS Marrows are no longer grown on compost heaps as used to be the custom, but in specially prepared plots. For this it is necessary to dig a shallow trench about 4 ft wide and into this put some 1–1½ ft in depth of half-rotted

28. *Marrows: Various bush types, including the round variety Tender and True.*

manure or well-made compost, or perhaps a mixture of equal parts manure and leaves, and cover with the soil which was taken out. This method should produce sufficient humus and fertility to grow a fine crop of fruits over a long period. On the other hand, a poor, hungry soil, with little or no feeding, will result in a very light crop and the plants giving up quickly and going off bloom as soon as hot, dry conditions occur.

A good time to make a Marrow bed outdoors is mid-May, and in a few days the plants, grown from seeds which have been sown under glass, can be put out. These young plants should be covered with cloches to avoid damage by late spring frosts.

On the other hand, if plants are not available, sow seeds in patches of two or three in the bed, and cover with inverted flower-pots, with a piece of tile to stop up the hole. This method should quicken the germination. In due course, the pots may be taken off in the day and put back at night, and the hole can be left open for a little air. During really bad weather the pots should remain over the plants all day. The final distance between the plants should be $2\frac{1}{2}$ ft square if Bush varieties are grown, and each clump of two or three plants should be reduced to one when the plants are of a good size.

PESTS AND DISEASES   A careful watch should be kept for slug attacks; lime or soot can discourage these pests, and slug pellets will reduce their numbers or eradicate them altogether.

One of the chief troubles in most seasons with the cultivation of Marrow is virus disease, and it should be better known by gardeners. The symptoms to look for are the mottling of the foliage in light and dark green yellowish areas, the leaves being small and stunted, and often somewhat puckered. It is most important to be able to recognize this disease in young plants; in older plants the first leaves may be healthy, and the trouble occurs quite late and only on the newly-formed leaves. When virus shows in young plants, do not hesitate to destroy them at once. Nearly always the cause of this Mosaic or Virus is Aphid attack, and it is always advisable to spray with a systemic insecticide in the early stages of growth, or, if more convenient, dust the young plants with Gamma B.H.C. Dust or Malathion Dust, applied as soon as the damage or pest is seen and before the leaves commence to curl.

EXHIBITION   In most schedules, the class is for a pair of Marrows of table quality, not exceeding 12 in in length. Two young shapely fruits, well matched and nearly always green-striped, are usually the prizewinners. Ten points can be awarded, six of these for condition and four for uniformity.

Vegetable Spaghetti   This is a very easily grown vegetable which produces a number of Marrow like fruits some 8–10 in. long on trailing type of plants.

Cultivation of this subject is the same as for growing trailing varieties of Marrows.

Mature fruits should be boiled for about twenty minutes and the result is just like spaghetti and may be eaten as a salad, seasoned with salt, pepper or butter. Can be served with spaghetti sauce for a delightful meal.

Courgettes   Courgettes are baby Marrows cut when the fruits have grown 4 or 5 in. long and they are cooked whole.

Usually types of Green Bush Marrows are used to produce Courgettes and for the most prolific and early producing crops it is advisable to use the F_1 Hybrid variety.

The fruits of this are of medium dark green colour of skin and for the best crop it is essential for the site to be deeply dug and the soil well fed. By this means a long continuity of cropping will be ensured and it should be noted that frequent picking is necessary to help maintain a continuous setting of young fruit.

Melons

Cucumis melo

ANNUAL

In horticultural circles, this subject is classified as a fruit; nevertheless, it would be remiss if it was not included in this book, for the reason that the seed of Melon is offered in the vegetable section of all seedsmen's catalogues.

With each successive year, more and more amateur growers are installing greenhouses, some heated, others cold, some well equipped with cold frames or Dutch lights, and, indeed, cloches. Whichever form of glass protection is available, there is a variety to suit it; and to be of success it is very important to grow only the variety most suited to the form of protection available.

HEATED GREENHOUSE Without doubt the finest Melons in the world are the English hothouse varieties; even personal taste is catered for: colour of flesh (in addition to flavour) is available in Scarlet, Green, and White—Superlative (Scarlet), Ringleader (Green), Hero of Lockinge (White). Seed may be sown at any time between January and mid-May in 3-in. pots, filled with J.I.1 for preference. Seeds are best sown by pressing edgeways into the compost about $\frac{1}{4}$ to $\frac{1}{2}$ in. deep. After sowing, place in a warm position; the seedlings should be showing in the course of a few days. These should be kept growing steadily, but care should be taken to place the plants as near the glass as possible to avoid their becoming drawn.

When five or six leaves have formed, the plants should be transferred to permanent quarters, which can be in the greenhouse border or in 9-in. pots or boxes. Unquestionably, planting in the border of the house is the most satisfactory method of cultivation. In addition, the position of the bed is very important. The soil should be firm; you can ensure this by using a good turfy loam, placed on the top of the manure. It is an excellent practice to make up the beds several days before planting, as this permits the soil to warm up. Avoid burying the roots too deeply; keep them as near the surface as possible, making quite sure that the stem of the plant is not covered with soil, as this is one method of avoiding canker. Spacing in the beds should be $2\frac{1}{2}$ to 3 ft apart. The growing point should be pinched out immediately above the fifth leaf to encourage lateral growth, and the growing point on the laterals should be pinched out in a similar manner. It is on these laterals that the male and female flowers will develop. (See drawing for the difference between male and female flowers.) Having identified the females, these will require artificial pollination; this is carried out by taking off the male bloom, stripping the petals and pushing the exposed centre into the female flower, where it should be left. This operation is best carried out at midday if possible. It is advisable to endeavour to treat all the female flowers on the one plant at the same time, thus ensuring a uniform crop maturity. As soon as the young fruitlets begin to swell, it is advisable to thin out to three fruits per plant, with a maximum of four.

In seed-sowing and the raising of young plants one should aim at a minimum temperature of 70°F, and, indeed, the same temperature should be maintained throughout the life of the plant. Feeding with a liquid manure should be commenced when the fruits are about the size of an Orange, and should be stopped when ripening commences. At the same time, watering the plants should be slightly reduced but not too drastically as the hall mark of a good Melon is a healthy foliage when the fruits are ready for cutting.

One of the disappointments experienced by

the amateur in producing this subject is due to failure to shade the glass against the direct rays of the sun, which may cause scorching of the foliage, and on a hot day to insufficient ventilation around midday.

FRAME CULTURE   If growing in a frame, it is best to use a three-light frame, as the heat will be more constant than with one of smaller size. Begin about the middle of April to prepare the bed. The best soil for Melons is a firm, turfy loam, 9 in. of which should be placed on top of the manure. It is a good practice to raise the plants in pots and have them strong enough to plant out as soon as the beds have settled down to a steady temperature. If plants cannot be raised in advance, seed can be sown on the bed. A sufficient number of seeds should be sown to provide for all contingencies.

The work of syringing, ventilating, and watering must be guided by weather conditions, and only done when the temperature is warm. The plant should never be dry at the root, and must have a light shower over the leafage twice a day. As the flowers open, the watering at the root should be discontinued, and the syringe should be used only in the evening when closing down the lights. In frames, Melons do better spread out on the beds, and each fruit must be supported by a flat tile or an inverted flower-pot.

When fruits are as large as the top joint of a man's thumb, watering should be resumed, and the syringe used twice a day until the fruit begins to change colour. Under this method of cultivation it is possible to use the Hero of Lockinge (one of the heated greenhouse varieties), but Ogen, Charantias, and Dutch Net are the most popular for this method of cultivation. A timely reminder; do not forget to shade the glass to prevent Leaf Scorch.

CLOCHE CULTIVATION   In recent years, particularly in the southern part of the country, this has become a very popular method, but it is most important to attempt to grow only those varieties such as Sweetheart and Charantias, both of which are quick-maturing. Further, it is a sound practice to be content with three fruits per plant, which will ensure a useful size of fruit, making the effort in production worthwhile.

The question is frequently asked: 'Is it necessary to pollinate by hand when grown by this method?' The answer is 'Yes': by so doing you are certain that pollination has taken place; whereas if left to the bees and other insects there is always a risk of sterility.

29. Melon: Superlative. Without doubt the king of the hot house Melons; scarlet flesh of exquisite flavour.

Mercury or Good King Henry

Chenopodium bonus—henricus

PERENNIAL

Mercury is a native of this country and for years has been grown in Lincolnshire and used as a substitute for Spinach.

The soil for Mercury should be well trenched and in a good state of fertility; this crop does best in a dry sunny position.

Seeds should be sown during April—1 in. deep in drills about 12 in. apart. Thin out the seedlings in due course to 9 in. apart. It is, however, possible to sow in a seed-bed and transplant to permanent quarters.

This is a very easily-grown vegetable, and, if placed in a well-manured, warm, sheltered spot, produces an abundance of tender shoots, which may be blanched. Under favourable conditions it can be cut a fortnight before Asparagus in the spring. The young shoots can attain the size of one's little finger.

Mushroom

Psalliota campestris

Most people regard the Mushroom as a vegetable and it is of course used as such; but it is in fact a fungus, and in many ways is most unlike any other vegetable crop.

There is always a large demand for Mushrooms throughout the year and, in consequence, shops can usually supply an abundance; but it is certainly well within the capacity of the amateur gardener to produce first-class buttons for a considerable part of the year.

But first the principles of Mushroom culture should be thoroughly understood. They can be successfully applied in many different ways, rendering the practical work easy and tolerably certain.

Mushrooms may be grown in cellars, barns, sheds, frames and greenhouses, using beds or wooden trays, or outdoors in ridged beds. Where an inside site is chosen, it should be free from cold draughts and perfectly clean. Special care must be given to surrounding timber or wooden structures where insect pests could be harboured. The best procedure is to treat the wood on the outside with a preservative such as creosote and, when the beds are again made up, the woodwork should be taken apart and reversed, so that the inside surfaces are on the outside, and these in turn given a dressing of wood preservative.

Stable Manure is the ideal medium in which to grow mushrooms. It should be heaped on a fairly wide base to enable it to generate heat, and after about a week it should be turned, making quite certain that the material on the outside is placed on the inside of the heap. This process should be repeated three or four times at intervals of three days after the first turning. The manure is fit for making into beds when it is free from objectionable smell, has a good brown colour, and is short in texture. On applying pressure, moisture should ooze from the material and the straw should break fairly easily.

Manure containing wood shavings or sawdust should be avoided.

MAKING THE BED   The bed can be made up to any dimensions, but the depth should be between 10 and 12 inches, and the manure should be pressed down firmly before spawning.

SPAWNING   Spawning of beds should not commence until the temperature of the manure has fallen to 75°F. Cast the spawn over the surface of the compost and ruffle it in as deep as you can with your fingers. Press back the compost and consolidate it firmly in the bed. A quart of Sutton's Al Spawn should spawn from 40 to 50 square feet of bed. The temperature of

manure should never be allowed to drop below 65°F before spawning.

CASING   After about ten days, the spawn will begin to 'run', and may be seen as a blue-grey mould if portions of the surface layer are carefully lifted. At this stage, the beds should be cased with about an inch of soil. Special care and attention should be given to the selection of this casing so that it will be free from weed-seeds, fungi, and bacteria—in other words, sterile soil. The heavier this soil the better, providing it is dug during dry weather and can easily be broken down. If clay is used, carbonate of lime and sand could be added to make it more workable in preparing this casing.

Mushrooms should appear eight to ten weeks

30. Mushroom: the heavy cropping white-capped variety.

after spawning, and beds should continue bearing for three or four months. Mushrooms should be pulled and not cut. The air temperature during cropping should not exceed 60°F.

OUTDOOR CULTURE   Mushroom spawn may also be planted in lawns or pastures in May or June, when there is a reasonable chance of obtaining mushrooms in August and September, provided that weather conditions are favourable. Heavy rain after planting, or a long spell of dry weather, may kill the spawn.

Mushrooms are not generally successful in poorly-drained land or in very dry, sandy soils. An open position, unshaded by trees or hedges, should be chosen.

Spawning is done as follows—a triangular tongue of turf, 2 in. thick, is raised and a piece of spawn inserted underneath so that it is in good contact with the surrounding soil. The turf is then pressed back in position and trodden down firmly. If good compost or weathered manure is available, a quantity placed in with the spawn will assist matters.

Mustard and Cress

Brassica alba and B. Nigra;
Lepidium sativum

ANNUAL

A native of Europe, Mustard is widely used as a pungent salad and is often used with Cress. As the two subjects grow at a different pace, they must be produced in separate containers.

Mustard is grown and cut when in the seed-leaf stage of development. Mustard should be sown three days after the Cress so that they are fit to cut together.

The best kind for salads is the common White Mustard (*Brassica alba*), as, although seed of black or brown Mustard is sometimes used, most people consider the flavour is somewhat too hot. On the other hand, a great deal of the so-called Mustard sold for salads by greengrocers, etc. is grown from Rape seed (*Brassica napus*) which cannot be distinguished from Mustard in the seed-leaf stage of growth. Rape is most commonly grown as an agricultural plant in this country, covering hundreds of acres, and is the source of colza oil, used for the manufacture of oil-cake for cattle-feed.

The most important factor when growing either Mustard or Cress is to obtain fresh

vigorous seed of the highest quality and germination. You should have seed which will germinate very evenly and continue to develop at a uniform rate. Therefore, do not keep Mustard and Cress seed for any length of time, and certainly not from one season to the next.

The best method to grow this salad in a small way is in a shallow box or tray in a greenhouse or frame, or put the boxes outdoors under cloches so that the leaves are quite free of any splashed soil, as would be the case if grown outdoors without covering. The ideal temperature for this work is a maximum of 65°F in the day and a minimum of 50°F at night. The same tray of soil can be used for two crops. The quantity of seed used to sow on light soil in an ordinary seed tray is $\frac{3}{4}$ oz of Mustard and $\frac{1}{2}$ oz of Cress; and, because they germinate better in the darkness, cover with light-proof paper, boards, or mats. This covering should be removed when the seedlings are 2 in. high and the seed-leaves are throwing off their seed-coats. At this stage, it should be finished in the light so that it will develop a good green colour. Generally speaking, the crop should be ready to cut in about ten days in the spring, and fourteen days during the winter.

For many years home-produced Mustard and Cress has been sown on flannel which has been kept moist in a warm room; it can be grown in this way at any time of the year and obviously produces the cleanest possible crop.

The ordinary Garden Cress has been grown for salads in this country since the sixteenth century, but is a native of Persia. It does not grow wild in England. It is grown and cut in the seed-leaf stage of development for use in salads with White Mustard or Rape, which is grown separately but in exactly the same manner. That is to say, as the two subjects grow at a different rate, the Mustard and Rape should be sown three days after the Cress so that they are fit to cut at the same time.

The crop can be cut when colour in leaf is attained by using a pair of large scissors in the right hand and holding the tops of the Cress in the left hand. Make the cut about a quarter of an inch above the soil and take care not to pull the roots up or disturb the soil.

Okra

Hibiscus esculentus

ANNUAL

Okra is a native of Africa and is largely grown in the West Indies under the name of Gumbo; also in the southern part of the United States, and in southern France. In warm climates it grows to some $4\frac{1}{2}$ ft high and produces pale yellow flowers, some 2 in. in diameter, with purplish red centres. These produce tapering seed-pods borne in an upright manner in the joints of the leaves. The pods grow quickly and in hot weather will soon get tough. They should be used within a week when they reach maturity.

This plant requires a long growing season and usually takes about five months from time of sowing to maturity. Seed should be sown not later than the end of March under glass in a temperature of 60°F and potted on into 5-in. pots, using John Innes Compost. Warm growing conditions are necessary, and it is only under the most favourable conditions that this will succeed outdoors in this country.

The plants will grow to about 15 in. high and produce four or five curiously-shaped pods.

Onions

Allium cepa

BIENNIAL

Very few kitchen gardens in this country would be complete unless it contained its Onion crop, which clearly shows its importance in everyday dishes provided at table in one form or another for flavouring a separate dish or as a component in cooking, which indicates over how many years the population of this country has realized its value. Its origin is not definite but history shows that it was extensively grown by the Egyptians, Greeks, and Romans. Onions were grown largely in England in Queen Elizabeth's time, and Bradby, writing in 1718, said that it was grown more extensively in the garden than any other root.

The amateur grower of today had the choice of several methods in the cultivation and production of a useful crop, but before entering into details it would be well to list them:

1 Sowing *in situ* in the spring and thinning out to the required distance, according to the size of the bulb one wishes to harvest and store.
2 Sowing in seed-trays in a greenhouse in January or February and transplanting to the open ground in April.
3 Sowing in a seed-bed in August a variety less prone to bolting, transplanting in February or March to final quarters.

The most recent introduction: sowing seed *in situ* at the end of August, singling in the early spring for harvesting in June/July.

5 Planting Onion sets in February/March.

It should be remembered this list does not cover the cultivation of exhibition specimens, which is dealt with separately in this chapter.

CULTIVATION

Method No. 1—sowing *in situ* Practically all types of soil are suitable for this purpose, provided one golden rule is followed: good, deep digging and thorough manuring of the land during the winter months, especially where very heavy soils, such as clay and heavy loam, are involved. The earlier the work is carried out the easier it will be to prepare the land for seed-sowing in the spring, as the winter frosts will pulverize the ground and make it much easier to break down. The use of farmyard or stable manure is excellent, but, if this is not available or in short supply, the use of the compost-heap will assist in improving the physical condition of the soil. Where the grower knows from experience that he is dealing with hungry ground, a dressing of a complete fertilizer, such as Growmore, worked into the soil prior to sowing, at the rate of 2 to 3 ounces per square yard, will be most beneficial. Many amateur growers, particularly beginners, lack experience in the use of artificial fertilizers, but a useful hint is to guard against the excessive use of nitrogen, as this will only create a soft, lush growth. This produces strong tops but induces bull neck (or stiff neck) and often leads to disappointment after growing the crop, for, after harvesting and storing, the bulbs commence to be affected with neck rot, which eventually destroys them. March and early April, to take the country as a whole, is the ideal sowing time. The drills should be $1-1\frac{1}{2}$ in. deep and the rows 12 in. apart. Onion seed is one of the slow-germinating subjects even at the best

31. (Opposite) Onion: The Sutton Globe. A hard-fleshed typically English Onion with very fine neck suitable for spring sowing; first-class keeping qualities.

of times and, if you sow *in situ,* a few indigenous weed-seeds will frequently germinate long before the Onion seedlings can be seen. There-fore it s a sound practice to mix a small quantity of Radish seed with the Onion seed, because, as Radish germinates very quickly, it will mark the rows long before the Onion seeds emerge, enabling the grower to run the Dutch hoe through the ground, thus saving unnecessary hand-weeding at a later stage. Thinning distances depend on the size of bulb required. However, well-grown bulbs of about 2 to 3 ounces, correctly harvested and stored, are the longest keeping, so a singling of about 2 in. is the ideal distance; but if a larger size is required, the singling distance should be increased accordingly.

If the grower knows that the soil at his disposal contains a legacy of weed-seeds, then the best advice which can be given to avoid endless work and possible disappointment is not to use Method 1, but to adopt one of the other methods of production.

Method No. 2 Where the grower is the proud possessor of a heated greenhouse, he can easily raise plants for transplanting to the open ground, which will go a long way to overcoming the question of indigenous weeds mentioned previously. All that is necessary is to obtain some ordinary seed-trays, fill these with John Innes No. 1, thinly broadcast the seed over the surface, lightly covering with the same soil, keep well watered, cover the trays, and bring them along in a temperature of about 50°F. A few days before transplanting to the open ground in their final quarters, stand the trays out in a sheltered spot in the open to harden the plants off. Row distances and spacings in the rows are again dependent on the size required, but, by using this method, larger bulbs can be produced, provided they are given the wider spacing.

Method No. 3 Sowing in August in a seed-

bed—the earlier part of the month in the north and the middle of the month in the south—and select a variety less prone to bolting, such as Sutton's Solidity, and left until February/March. Finally, transplant to permanent quarters, according to climatic conditions, in a similar manner to that recommended in Method 2. It is always good practice that, if the bed contains a few weeds say by October/November, leave them in the bed, thus affording a little protection in the event of bad weather during the winter. One of the advantages of August sowing is that the bulbs will mature and ripen two or three weeks earlier than from either sowing *in situ* as in Method 1 or from a greenhouse sowing and transplanting as in Method 2.

Method No. 4 As the result of extensive breeding, new varieties of Onions of a very distinct type are now with us, and by sowing *in situ* in the last week of August and thinning to the required distance in February/March, varieties such as Express Yellow F_1 Hybrid and Kaizuka Extra Early can be raised. These mature and are fit for harvesting in late June when English stored Onions from the previous year's crop have long since finished. This overcomes the necessity of purchasing foreign produce which hitherto have been necessary to cover the period of late June to early August for household use. It is pleasing to note also that these new varieties have very fine necks which assist quick ripening, so beneficial in storing and long keeping.

Method No. 5—Growing from Sets If it were possible to talk to the professional gardeners of thirty or forty years ago, few, if any, would have knowledge or practical experience of this method of cultivation, which shows how rapidly this method has come to the fore. By using a variety such as the Stuttgarter Giant, it is the easy way for the amateur to grow a crop, particularly on land where indigenuous weed-seeds can make hard work of cleaning the bed. The land is prepared as previously mentioned for other methods, and the sets are planted as soon as conditions permit in February or March. Methods of planting can vary according to types of soil, but the easiest approach is to put the line across the garden, draw a shallow V-shaped drill, place the sets in and lightly cover; some growers, provided they have a good tilth, just press the sets into the soil.

EXHIBITION   This subject is, of course, the pride and joy of every keen exhibitor. To walk around a show bench and to listen to the grower who has succeeded in producing a bulb of 2 to 3 lb is really an experience. Nevertheless, it is safe to say that the individual who has accomplished this, will seldom, if ever, divulge all his secrets. In spite of this, provided the beginner is prepared to work hard on a long-term basis, there is no reason why he should not eventually accomplish what others have done before him.

The physical condition of the soil where large Onions are successfully produced is not entirely built up in one year. Some exhibitors have grown exhibition Onions on the same piece of ground for a number of years, continually building up the standard of fertility. The first essential is trenching the land, which really means 3-spits deep, incorporating well-made farmyard manure. Break up the sub-soil in the bottom of the trench, place 5 or 6 in. of manure, cover this with the next spit of soil, covering it with another layer of short manure, and finally replace the top spit. On no account must sub-soil be brought to the surface. This trenching should be carried out as early as possible in the winter months, leaving the surface of the soil as rough as possible, thus allowing the frosts to assist in breaking down the surface and making it in a friable condition by planting time. On very heavy soils it may be advisable to double-dig the ground in preference to trenching.

Sowing the Seed   It is essential to commence this operation in the last days of December or the first few days of January, using a first-class reliable strain, such as Sutton's Selected Ailsa Craig or a similar variety. Use either a flower-pot or a seed-tray filled with John Innes No. 1 seed-sowing compost, lightly covering the seed, and maintaining a temperature in the greenhouse of about 50°F. As soon as the seedlings are large enough to handle, they should be pricked out either into single pots or into seed-trays which have been filled with John Innes No. 3, or, if you prefer to mix your own, 2 parts fibrous loam, 1 part of spent mushroom-bed, adding sufficient sharp sand to keep the whole porous. If making your own compost, it is advisable to do this during the winter, keeping it under

32. Onion: Suttons Selected Ailsa Craig, a large growing onion particularly suited to early sowing under glass for the production of fine exhibition bulbs.

cover, and before use, it should be passed through a $\frac{1}{4}$-in. sieve. The plants should be ready for transplanting to the final quarters about mid-April, care being taken to harden off the plants thoroughly before transplanting. If, after transplanting, conditions are dry due to April winds, do not hesitate to water to prevent the plants flagging. As soon as they are well established, a dusting of artificial manure (such as Clay's) should be applied every ten days. Be generous in your spacings, the ideal distance being 2 ft between the rows and 15 in. from plant to plant. It will be appreciated that, to obtain a large bulb, it is necessary first to get a sturdy, well-grown plant; hence the regular feed already mentioned—but care should be taken not to overdo the nitrogen, as, here again, this could lead to bull necks, which are so undesirable.

Preparing and Staging The majority of the shows throughout the country are held in the months of August and September; it is therefore no good waiting until the day before the show to lift the requisite number of bulbs for exhibition. Apart from the time factor, why go to the trouble of raising first-class bulbs if they are badly prepared and exhibited? The earlier the date of the show in the season, the earlier you must commence visual observation of the growing bulbs. It is quite a simple matter to mark a number of the bulbs with short sticks as being the most uniform in size, shape, and colour. Immediately a decision is taken as to which bulbs are required, carefully ease the soil away from the base of the bulb to allow the light to colour the skin evenly or ripen at ground level. At the same time, loose skins which may be present should be carefully removed, again to allow even colouring and ripening. The bulbs can then be lifted and exhibited in good condition.

For late July and August shows, one could not expect fully matured and ripened large bulbs. It is therefore a good practice to lift and

stage, sometimes complete with foliage; or perhaps the most attractive method is to cut the foliage halfway down, forming a fan shape, and, if six bulbs are required by the schedule, place each in a small flower-pot to hold them rigid in the formation of 3–2–1 from back to front. For September shows onwards, the bulbs should be lifted ten to fourteen days prior to the show, surplus foliage removed, and the necks tied in neatly, as close to the neck of the bulb as possible, when they have reached a nice ripened colour. At certain times and in certain seasons, a stronger neck of the bulb will make it difficult to tie down, this can be overcome by running the blade of a pocket-knife carefully into the back of the neck, removing some of the interior, and then tying down. The reason for advising lifting and preparing well in advance of show date is to enable the grower to place the bulbs in full sunlight to produce an evenly coloured outside skin, which is so attractive on the show-bench. Fight at all costs the temptation to remove layers of skin, as this will lose valuable points when judged.

STORING   The art of storing Onions is to provide a cool temperature with plenty of fresh air. If facilities are available, nothing is better than wire netting fastened to a fixture to allow the air to pass below and above the bulbs. Another method, of course, is to tie in bundles or ropes and hang in a very cool place, again allowing air to circulate all round.

DISEASE AND INSECT PESTS
White Rot   This disease, which is caused by a fungus, is fairly common. The earliest signs of attack are often seen towards the end of May in plants from an autumn sowing. The older leaves turn yellow and then fall over, and affected plants can be easily pulled from the soil because the roots are rotten. Around the bottom of an attacked bulb a white fluffy growth develops, destroys the roots, and attacks the bulb. All such bulbs must be destroyed at once as the

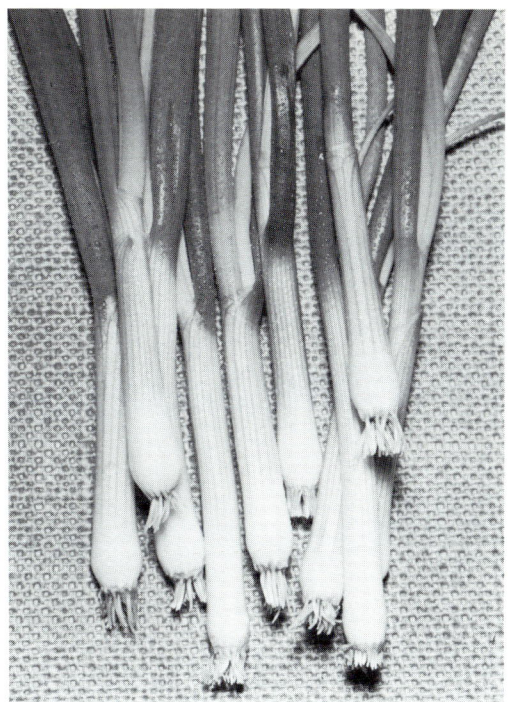

33. Onion: White Lisbon. A mild-flavoured silver-skinned Onion for quick salad production only.

soil becomes contaminated. Obviously it is advisable to rest this piece of ground for several years, and not sow or plant any member of the Allium family on it during this period.

A fair measure of control can be obtained by using a 4% Calomel Dust in the seed drills at the time of sowing. For ease of application, mix the Calomel Dust with sand or dry soil. A 25-yd length of drill requires approximately 1-lb of Calomel Dust.

A number of Onion varieties show some resistance to White Rot, but the salad Onion, White Lisbon, and some stocks of Ailsa Craig are rather susceptible to this disease.

Mildew In some seasons, when the humidity of the atmosphere is high, this can be troublesome if immediate action is not taken. As soon as it is seen the plants should be dusted with Bordeaux Dust.

Neck Rot Has already been mentioned in a previous chapter.

Onion Fly To prevent an attack of this, it is advisable to dust the drill with Calomel Dust before sowing.

COCKTAIL ONIONS Most people enjoy Cocktail Onions pickled in white vinegar. This type of Onion can also be used in making piccalilli, cooked as a separate dish, or cooked with Peas and young Carrots in the French method. Preparation of the soil is similar to that for an ordinary Onion crop, but it should be remembered that these are very quick-growing and that, in general terms, they will not store and keep in their natural form. The best method of approach to this subject therefore is to make successional sowings, the first in March for June/July use, the second in May for August/September, and the last sowing in early July for autumn use.

Onion, Potato or Underground

Allium aggregatum

PERENNIAL

This type of Onion is not grown from seed but is perpetuated by replanting the bulbs in exactly the same way as is done with Shallots (*q.v.*).

Planting is best done in February or March according to weather and soil conditions. Plant the bulbs 9 in. apart with 1 ft between the rows; place the bulb at least 1 in. below the soil.

To obtain good-sized bulbs dig in some well-rotted manure some time before planting.

The crop is ready for harvesting in August.

Onion, Tree (or Egyptian Tree Onion)

Allium cepa var. proliferum and *var. viviparum*

HARDY PERENNIAL

This subject cannot be grown from seed and it is therefore propagated by planting the bulblets in the late autumn months. The bulbs are rather flat and coppery in colour.

Under average conditions this subject grows to about $2\frac{1}{2}$ to 3 feet. As the bulblet develops into a plant with Onion foliage, it throws up, in the spring and early summer months, a centre pipe, much like an ordinary Onion seed stem, but, instead of developing a flower and seed-head, it produces a cluster of small bulblets which can be used for flavouring or for ordinary culinary purposes.

Plant the sets about 1 in. deep and 9 in. apart, and, as with Shallots (*q.v.*), make quite sure they are firm, to discourage birds from pulling them.

Onion, Welsh

Allium fistulosum

PERENNIAL

A native of Siberia and the East, this subject can be cultivated as an annual or biennial but botanically is a perennial. It never forms a rounded bulb but consists of numerous long, white-based scallions, almost resembling small Leeks. During World War II it was widely grown as Bunched Onions or Spring Onions when the popular White Lisbon was in very short supply.

There are in fact three types of Welsh Onion, two of which are grown from seed—one with a white stem, the other coppery red—but the most popular type is the perennial form which does not produce seed but is perpetuated by the sub-division of the parent plant. It is extremely hardy.

Usually with this type of Welsh Onion it is best to divide the plants every two or three years, retaining only the youngest portion; in this way, new growths which look fresh and green are always available.

Orache or Mountain Spinach

Atriplex hortensis

ANNUAL

Originally a native of Tartary, Orache produces a plant with broad, arrow-shaped, pliable leaves, and is sometimes grown as a substitute for Spinach. It grows to a height of 5 or 6 feet. Seed may be sown in drills 2 ft apart in the open ground from early March onwards. Orache is not suited to transplanting.

As soon as the seedlings have made three or four leaves, they should be thinned out to about 18 in. apart. An occasional watering is advised in dry conditions as the plants are inclined to run to seed otherwise. In such circumstances, successional sowings are advised.

In gathering Orache, it will be found that the young, tender leaves are decidedly better-flavoured than the larger, older ones.

Parsley

Petroselinum crispum

BIENNIAL

Parsley was grown on the continent for several hundred years before it reached England, but even so this plant has been grown and used here since about the year 1500; and Henry VIII is said to have enjoyed Parsley sauce with roast rabbit.

SOILS   Most well-drained soils will grow good Parsley, but to produce a really first-class crop a well-manured, moist soil is required. It succeeds well on a medium-to-heavy soil, but even the lighter soils, if in good fettle, will yield a fine cutting for many months.

99

TIME OF SOWING Seed can be sown outdoors from March onwards, when soil and weather conditions are favourable, for cutting or picking in the summer months. Sow during the month of June for winter cropping and early in August for the following spring. Where, however, there are greenhouse facilities, sow a little seed in gentle heat in February. Prick off the seedlings into trays or boxes and transfer to the open ground towards the end of April in rows 12 in. apart with 6 in. between the plants.

Parsley seed germinates slowly. It takes five weeks and more before the seedlings show through the soil, and some gardeners mix a little Lettuce or Radish seed with the Parsley as either of these will germinate quickly and mark the rows so that hoeing can be done before the surface of the soil becomes covered with tiny seedling weeds. It will also assist the early thinning of the Parsley rows.

Parsley transplants readily, either from trays sown early under glass or from the rows sown *in situ* at thinning time. It is a plant which needs ample space for full development and on no account should the rows be left unthinned, as the individual plants and stems will be small and poor.

Parsley is frequently sown along the edge of the garden path for ease of picking and also for its attractive appearance. Try to pick the rows constantly—take a leaf or two from each plant and do not strip any one plant.

It is quite a simple operation to lift a few plants in the autumn, pot them in 5- in. pots, keep them in a frame for a few weeks, and then put in the greenhouse to provide fresh green leaves during the winter.

Two of the best varieties are Curly Top and Claudia D.4, both having fine long stems, closely-curled foliage of a dense, dark green colour; both are hardy in character.

Parsley is used both for garnishing and seasoning and there is little doubt that it is one of our most popular Herbs.

The leaves when dried and rubbed into a powder can be stored in tightly-corked bottles for use in the winter months for seasoning, etc. Naturally, the best Parsley for drying is the young leaves, taken in the summer months. (See also HERBS.)

Parsley, Hamburg or Turnip-rooted

Petroselinum crispum

BIENNIAL

Hamburg Parsley is quite distinct from the ordinary garnishing type, and is grown for the sake of its small Parsnip-like roots, which are widely used on the continent. The roots really resemble those of the Parsnip, and the flavour has often been described as a combination of Celery and Parsley.

The cultivation of Hamburg Parsley is very straightforward; the seeds are sown in drills in April, spaced 15 in. apart. As with ordinary Parsley, the germination is rather slow and it is a good idea to mix a little Radish or Lettuce seed with the Hamburg to act as drill markers in the early stages of growth. Thin the seedlings when of suitable size to a distance of 9 in. apart.

Roots attain a useful size by the early autumn, and in November you can lift those which are left and store in a similar manner to Beet (*q.v.*).

Most soils are suitable for this crop, but, as with other root crops, freshly-manured soil is not to be recommended, as it induces forked roots.

Parsnip

Pastinaca sativa

BIENNIAL

Parsnips have been grown throughout Europe for some 2000 years, and have been grown and appreciated in Great Britain for several hundred years.

SOILS   Most soils will grow a worthwhile crop of Parsnips, except perhaps the thin, gravel soils which will only produce roots with small inferior fangs. Naturally, one of the best soils for this crop is the deep sandy type, provided it has been properly prepared by deep digging, ideally in the autumn. The introduction of fresh manure will most probably encourage fanged roots.

TIME OF SOWING   From early March and throughout April the soil can be levelled down and the seed sown, providing a good tilth can be obtained and a fine seed-bed made. Sow the seeds in drills 1 in. deep, with rows some 15 in. apart. Either sow the seeds thinly or drop them in twos or threes at a distance of 6 in. apart. Carefully cover with fine soil and rake lightly to leave a neat finish.

GROWING METHODS   Thin the seedlings when they are growing strongly and are well established, and finally leave the plants 12 in. apart; but, should the grower have a show in mind, increase this distance to 15 in.

Parsnips are usually left in the soil until they are required, but in the harder winters lift a portion of the crop in November and store under cover. It is a recognized fact that the flavour of the root improves if left in the soil until the end of February, but growth recommences after this and then they are best lifted and stored.

Perfect-quality roots for exhibition work require somewhat special cultural methods.

Use a crowbar to make really deep holes (2–3 ft deep), 7 or 8 in. across the top, and fill these with specially-prepared, fine potting soil. Make this quite firm in the holes and sow 2–3 seed per station, finally reducing to one plant only. At lifting time in the autumn the greatest care should be taken; it will be necessary to remove a few spits of soil alongside the rows to avoid damage to the long, perfectly-shaped taproots. R.H.S. points value, 20.

A recent introduction to the somewhat short list of varieties is Sutton's White Gem, which is ideally suited to the shallower soils and to those gardeners who prefer the shorter roots. In addition it shows a considerable tolerance to Canker.

However, Tender and True is still considered one of the best varieties for both general use and exhibition. It is a perfectly symmetrical, smooth, white-skinned and white-fleshed root of outstanding flavour, besides showing a marked resistance to Canker.

34. Parsnip: White Gem. An outstanding new variety for shallow soils and for those growers who prefer a shorter growing root is highly resistant to canker.

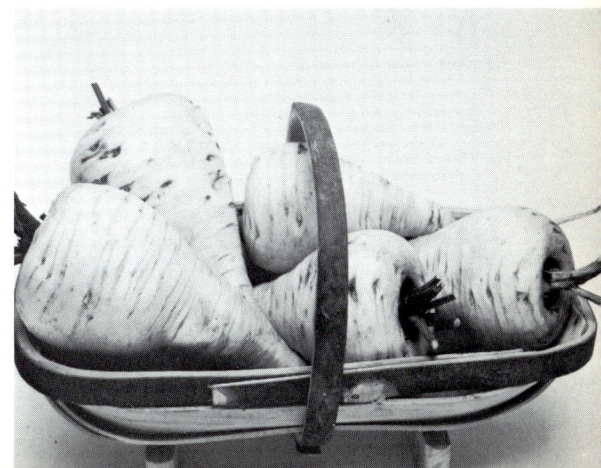

Peas

Pisum sativum

ANNUAL

Unquestionably, the Garden Pea is one of the most popular vegetables of all, not only as a variation from many others, but also because it can be the first dish of the season that really signals the arrival of the early summer months. It also creates a challenge to the grower, and can cause friendly rivalry in the grower's area to see who has produced the first dish of Peas of the season. Its history goes back many centuries. Peas were certainly grown by the Greeks and Romans, and, although it is recorded that it was first introduced into England from the continent during the reign of Henry VIII, England can safely claim the development, through breeding, of a wide range of varieties in height of plant, colour, shape, and size of pod, and, most important, in table quality. Further, the first wrinkled-seeded types were the result of breeding work in this country.

Many amateur growers fail to understand the difference in uses between the round and wrinkled varieties. The round-seeded varieties have a starch make-up; whereas the wrinkled contain sugar. That is the reason why only the round varieties should be sown outdoors, with or without cloche protection from November to the end of February, and, generally speaking, the wrinkled types from March onward, for the starch content of the round seeds indicates hardiness.

The huge list of varieties which were available thirty years ago have, due to economic conditions, been drastically reduced in number to meet prevailing conditions. The emphasis today is placed upon dwarf-growing varieties which serve several purposes. The need for costly, tall pea-sticks is done away with, and, as described later in this chapter, they can be grown with less space between the rows, thus saving land and obviously increasing the yield from a given area.

SOILS: PREPARATION AND CULTIVATION   This subject is adaptable to most types of soil from light to heavy and responds to good soil preparation. Particular attention should be paid to drainage on heavy soils, indeed, such land should receive a good coating of humus in some form or other, whether it be stable manure, farmyard manure, or from the compost-heap. Where this is not possible a good substitute would be a dressing of a complete fertilizer, such as Growmore, at the rate of 3 oz per square yard worked into the soil before drawing the drills.

If one referred to the old text-books, one would find that the flat type of drill, 4–6 in. wide and 2–3 in. deep, was the general practice, staggering the seeds at an inch or so apart in the drills. Whilst this was a sound practice with the tall varieties being universally grown at that time, for example Ne Plus Ultra (7 ft tall), or such varieties as Alderman and Duke of Albany, which necessitated at least 4 ft from row to row; but in these modern times, with the emphasis on dwarf varieties which are anything from 1 to 2 ft when fully matured, a different method could well be used, which is drawing a V-shaped drill 2–3 in. deep, and sowing in a single line, considerably reducing the space between the rows. For example, with Feltham First, Early Onward, Kelvedon Wonder, and Little Marvel, it is recommended that the rows should be 1 ft apart; as the plants develop they will support each other without the use of twigs or sticks, thus reducing the cost of crop production and at the same time reducing the area of ground occupied by the crop.

Commercial growers, when producing field

crops by the acre, either for the fresh market trade or the canners and freezers, all use this principle. It is important to remember, however, that, with the considerably narrower distance between the rows, the Dutch hoe should be run through the ground as soon as the seedlings emerge, to kill any indigenous weeds that have commenced growing and to aerate the soil, which is so beneficial to the young Pea plants.

Peas grow very quickly, and as they reach flowering stage the plants will prop each other up, and in so doing will exclude the light and smother any late weed seedlings which may appear.

VARIETIES As a result of many years' experience of crop yields of actual seed from a given area, it has been seen that some varieties give heavier yields than others and that some combine with yield superior table quality. The best of these varieties are Little Marvel, Kelvedon Wonder, Early Onward, Onward, and Senator, which can be relied upon to meet these requirements. All these varieties are dwarf-growing, with the exception of Senator, which on average soil would be $2\frac{1}{2}$–3 ft. For those who prefer and have the opportunity to grow the taller varieties, Suttons Sweetness is the earliest wrinkled seeded, height 3 ft., while Suttons Show Perfection ($4\frac{1}{2}$ ft.) is the thinnest shelled variety in cultivation, outstanding for exhibition and for freezing. Don't forget that, if you are growing these taller varieties, it is preferable to use the flat drill and allow 3 ft. minimum from row to row.

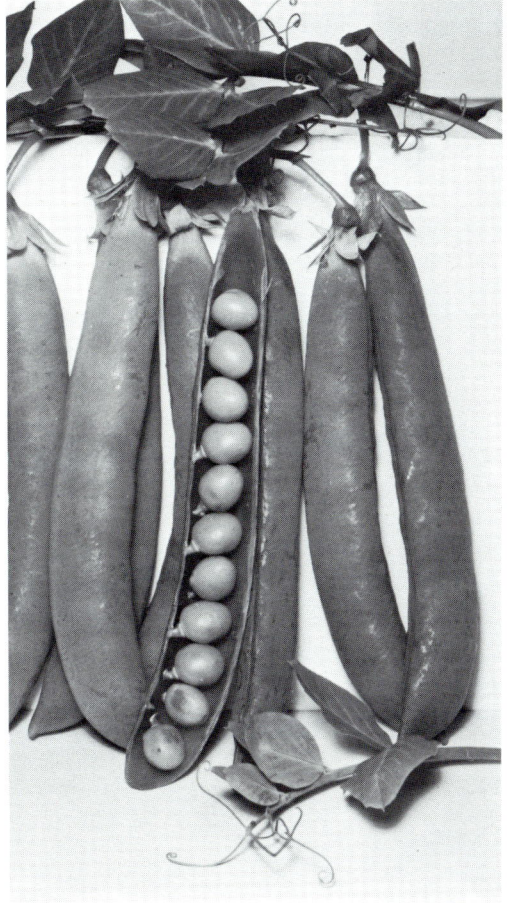

35. *Show Perfection: a first class thin-shelled pea, eminently suitable for the table as well as exhibition.*

EXHIBITION Many exhibitors gather their show material from a crop which was specifically grown for table use, but on reflection you will realize that it is a pure gamble that such material will be available for the date of the show. Therefore, the keen exhibitor will make a sowing specially to be in prime condition at the exact date, but again, as this subject is growing in a difficult period of the year when all types of weather are possible, the only way to be certain is to grow on the single-cordon system. With a variety like Show Perfection, for a show to be held at the end of July or in August, with long hours of daylight and possible dry conditions, it is wise to allow ten to eleven weeks for the date of sowing, counting backwards from the date of the show. Space the plants 6–8 in. apart, support them with bamboo canes, cut out all tendrils as they appear, and tie into the canes in a similar manner to that which is used for growing Sweet Peas for exhibition purposes. Allow 21 days from the actual date of the show before allowing any of the flowers to set. If, for example, there are thirty days to go, remove the first flush of flowers and allow the second bloom on each plant to set. By using this method, you can be almost certain of having first-class exhibition material on time. At shows, what might have been a first-class exhibit can be spoilt by rough handling; so stick to the following rules.

At no time should the actual pod be touched with the fingers, as every finger-mark removes a little of what is commonly referred to as the bloom on the pod. When gathering for show, have a box, well lined with tissue-paper, available at the side of the row, then cut the stem with a sharp knife or pair of scissors, holding the pod only by the stem. Leave a good length of stem attached to the pod, and lay in a single layer in the box. On arrival at the show, remove the pods from the box, again handling only by the stem; whether staging on a flat bench or on a dinner-plate, always put the stalks into the centre, making a circle of the pods.

Another common happening at shows is that a good dish of Peas fails to win a prize simply because the judges, on opening one or two pods, find that Pea Maggot is present, causing an immediate penalty. A very simple precautionary measure to prevent this trouble, is, when the flowers are opening, to spray with Murphy's Fentro. The same treatment may well be used also for crops grown for table use.

PESTS AND DISEASES   Generally speaking, this crop doesn't suffer unduly from diseases or insect pests, but one or two which come ´ to mind. Foot and Root Rot, for instance is usually found on heavy, low-lying land which is poorly drained. The crop grows to about flowering stage and then dies out. The disease is easily identifiable by pulling one or two plants; it will be found that from ground level to the roots they have gone completely black, and perished. Another common cause for disappointment is due to Thrips, which not only causes a silvering of the foliage but also distorts the head of the plant causing it to go fuzzy-headed. Immediately silvering of the foliage can be seen, spray with Liquid Malathion or dust with Malathion or Sevin Dust.

Potato

Solanum tuberosam Solanaceae

PERENNIAL

The Potato, a native of South America, was first introduced into Europe by the Spaniards in the sixteenth century and into England by Sir Walter Raleigh a little later.

A tremendous amount of plant breeding has taken place over the years with the Potato, and it has developed from the coarse, deep-eyed, misshapen tuber, as originally brought across the Atlantic Ocean, to the many perfect-shaped, shallow-eyed varieties of the present day.

However, this crop is still subject to a few diseases and it is well to have some knowledge of these and to know how to deal with them when they occur.

It is also important to secure a supply of seed which is both healthy and vigorous and suitable for the purpose you have in mind and also for the soil and district in which the crop is to be grown. Certain varieties show a decided preference for one type of soil and will only succeed in soil which is of heavy or medium texture and which retains the correct amount of moisture. On the other hand, some sorts succeed well on light dry soils and such types can be in great demand.

Every variety is listed under one of three groups: First Early, Second Early, and Maincrop. Most gardeners insist on a fresh seed supply each season of either Scottish or Irish grown seed, taken from crops inspected by officials of the Department of Agriculture, and conforming to their regulations, including immunity to Wart Disease.

Most types of soils can be made suitable for Potatoes, even the very heavy and clay types, which are improved by providing a good dressing of light, strawy manure, dug in to the land during the early part of the winter.

For successful Potato crops all soils should receive a liberal dressing of well-made manure or compost in the late autumn or early winter. When digging, make sure to leave the ground rough, to allow the frosts to help pulverize it in the spring.

Just before planting, a well-balanced complete fertilizer should be applied. The spacing of the sets when planting is often governed by the size of the garden, but it is obvious that nothing can be gained by planting too close. The Potato is a sun-loving plant and the foliage requires ample space in order that the full development of tubers can take place. Generally speaking, it is advisable to allow 18 in. for early varieties and a minimum of 24 in. for Maincrops, but where crops are to be lifted very early the

36. Potato: Foremost. One of the most successful Sutton bred varieties ever introduced. Heavy cropping, white skinned oval tubers.

distance can vary from 1 ft for Earlies to 15 in. for Maincrops. Obviously, slightly wider spacing, either between the sets or between the rows, will increase the weight of Potatoes from each root.

When planting, the top of the set should be about 4 in. below the surface of the soil, and all shoots except for two or three should be rubbed off.

TIME OF PLANTING   Under really favourable conditions it is possible to plant on a warm, dry site as early as February in sheltered spots, but, of course, such crops must be protected in some way in the event of severe weather conditions. Generally speaking, however, the latter part of March is early enough to plant crops out of doors. For Maincrop sorts, planting can usually take place from the end of March and through April, according to the district and the type of soil. It is always advisable to put off planting for a while if the ground is heavy and wet, as this can consolidate very quickly, making it impervious to the air and of no use for root penetration.

It is obvious that sprouted seed can be planted much later than the unsprouted tubers, without the risk of loss and valuable growing time.

Sometimes there can be a delay in planting beyond the usual period, and it is as well to realize that very useful crops can still be grown from plantings made in May, even towards the end of the month, but such ventures depend naturally on favourable weather conditions for success.

It is generally agreed that the dibber is not an ideal tool for Potato-planting as it can leave the tubers suspended in a pocket.

If the ground was manured in the autumn, it needs forking over again in the spring to produce a nice tilth ready for planting.

Quite a lot of gardeners like to draw a shallow drill and then plant the tubers at the required depth with a trowel; but, if you wish, the drill can be drawn to the full depth. Cover the tubers with earth and take care not to damage the tender sprouts.

As soon as the growth of the Potatoes appears, hoe the ground between the rows and if you suspect a frost the haulms should be lightly moulded over.

As growth continues, continue to earth up and always do this before the haulms develop too far, otherwise you may damage the foliage, which in turn will affect the crop. Earthing-up will prevent greening of the tubers through exposure to the air. It is also said to minimize the risk of blight spores coming in contact with the tubers.

On the light soils and also in dry areas it is best to leave the top of the earthing-up fairly flat, or the moisture will run to the base of the drills and the roots will not get through. However, in wet districts it is better to bring the ridges to a point.

For the prevention of Potato Blight or Potato Disease, it is imperative to spray the foliage with Bordeaux Mixture. Dark brown patches on the leaves, underneath which may be seen a whitish mould, particularly around the edges of the spots, means that the crop has been affected. When there are moist conditions, the disease can spread rapidly, and the spores that are washed down into the soil infect the tubers, producing dark areas on the skin and rusty spots in the flesh. Blighted tubers can also develop a wet rot.

It is imperative to commence to spray before the trouble starts, and, although the time of its appearances varies from season to season, it is usually during June in the west, July in the south and east, and slightly later in the north. It is as well to commence spraying ahead of these times and, should weather conditions favour the disease, one or two further applications of Bordeaux Mixture should be applied at intervals of two or three weeks.

Earthing-up should help in guarding the tubers from infection. Do not on any account leave Potatoes in the ground because the crop is not worth digging; and do not bury diseased haulms and tubers in a shallow trench thinking it is a safe way of getting rid of worthless material, as this will be a means of storing up Blight for another year. Bury such vegetation only at a considerable depth, but the best way, of course, is to destroy all Potato refuse by burning.

OTHER TROUBLES   The reason for blackening of the tubers when cooked can be difficult to determine, but it is often due to Potatoes being grown on soils deficient in potash, or where excessive nitrogen in relation to potash has been used. Frequently the use of a well-balanced potato fertilizer can put the matter right.

Blackleg Plants affected by this organism turn black at soil level, stop growing, and eventually collapse. As a rule, only occasional plants are affected and these are best dug out and destroyed. The disease frequently appears in June and does not spread from plant to plant although it can be carried on to future crops if diseased tubers are planted. As a rule, affected tubers decay quickly, but during a dry period the disease may be checked. Slightly-infected tubers, if stored, may continue to rot, and under damp conditions extensive decay may occur.

Common Scab This is most common in dry seasons, especially on light, hungry soils. Scabby spots form on the skin of the tubers and may cover the whole surface if diseased tubers are planted. As a rule, affected tubers decay quickly, but during a dry period the disease may be checked. Slightly-infected tubers, if stored, may continue to rot, and under damp conditions extensive decay may occur.

The scabs take various forms, some showing depressions, others round swellings, and still others appearing as small pimple-like growths. The trouble is only superficial and the scabs are easily removed when the Potatoes are peeled. It is known that liming the soil or the use of alkaline fertilizers encourages scab, and a slightly acid reaction is therefore an advantage.

Control is effected by the use of liberal quantities of organic materials, and a strict rotation to avoid too frequent planting of Potatoes on the same land. In no circumstances should scab-infected tubers be used for planting.

Dry Rot This fungus develops in stored Potatoes. The disease is usually first seen in December and tends to become worse as the tubers mature. Affected tubers should be destroyed at once, otherwise the trouble will possibly spread to adjoining ones.

Most early varieties are susceptible to the trouble, with the exception of Epicure and Home Guard. A few Maincrops will also sometimes show symptoms of the disease.

Hollow Heart This is a physiological disorder. Tubers appear perfectly sound but, when cut across, are found to have hollow centres. This condition usually occurs in seasons when wet periods alternate with very dry ones. Lack of organic matter and unbalanced fertilizing can accentuate this trouble. Hollow Heart is usually worse on light, dry soils, and some varieties have a greater tendency to behave in this way more than others. It is sometimes advisable to change the variety if Hollow Heart becomes troublesome.

Potato Root Eelworm This nematode is responsible for the condition often known as Potato Sickness, which occurs on ground that has carried Potatoes for several years. Patches of plants appear dwarfed and sickly and a poor crop is produced. These patches get larger in subsequent seasons and the ground becomes incapable of producing a healthy crop. Control of this pest is very difficult, and easily the best plan is to refrain from growing Potatoes on infested land. It is possible for Tomatoes to be attacked by this eelworm but any other plants can be safely planted. It is generally considered that about seven years is a sufficient period to elapse before again planting Potatoes. If it so happens that there is no alternative site for Potatoes and if a long-term rotation cannot be arranged, only Early Potatoes should be planted. By lifting these in good time, when the majority of cysts will be immature and still attached to the roots, the eelworm population can be greatly reduced, provided the plants are burnt at once. A satisfactory crop cannot be expected unless the ground is heavily manured.

Virus Diseases The Potato is subject to a number of disorders of this type, with Leaf Roll and Mosaic the most serious. Most virus diseases are transmitted in the main by Aphids, although rubbing together of the foliage may in some cases spread the virus from one plant to another. It is most important to purchase good quality seed Potatoes from a reliable source, and to make quite sure to have fresh seed each year.

Great attention is now being paid to the elimination of Virus diseases by the best seed-growers, and it is therefore important to purchase certified seed only. In the seed-growing areas of Scotland and Ireland the climate is comparatively wet and windy and, in consequence, the Aphid population is low; this is one of the reasons for the vigour and superiority of seed Potatoes from these sources.

Wart Disease Some years ago this disease threatened to become a serious menace, but the development of immune varieties and strict control of plantings on infected land have for all practical purposes eliminated it. It will be found that the majority of the popular varieties of the present time and certainly all the more recent introductions are immune to Wart Disease.

It must be explained that the term 'immune' applies only to Wart Disease; such varieties are not necessarily more resistant to Blight or any other disease causing tuber decay.

EXHIBITION Although it is sometimes poss-ible to find suitable tubers amongst crops grown for ordinary eating, the keen exhibitior will nearly always grow a limited number of sets expressly for the production of quality tubers for the exhibition-bench.

It is also possible to select amongst present-day varieties a useful number of varieties particularly suited to exhibition, and the following are among the best:

Di Vernon	early	kidney, white/purple eyes
Catriona	2nd early	long-oval to kidney, white/purple
Dr. McIntosh	early maincrop	long-oval to kidney smooth skinned, white
Red Craigs	2nd early	oval pink
Royal Maris Peer	2nd early	round/oval pale yellow

For such a purpose the tubers should be of a uniform size, but certainly not coarse. They should be shallow-eyed, clear skinned, and free from scab or other blemishes.

Seed should be set up in trays and not more than two eyes to develop; in fact a single sprout will sometimes help towards obtaining a good size. Surplus sprouts should be rubbed off as they appear.

The soil for this special plot naturally should be in good condition; a liberal application of well-decayed farmyard manure and leaf soil should be made, preferably in the winter or early spring.

If the soil shows a tendency to produce scabbed Potatoes, a good idea is to dig in grass-cuttings just before planting, and a sprinkling of soot will also be most beneficial. Fork the ground over in April and plant the sets in drills at 6 in. deep. A light dressing of a complete Potato fertilizer should be applied at planting time. Usually a little more space is allowed than for ordinary cropping; and give $1\frac{1}{2}$ ft between the sets and $2\frac{1}{2}$–3 ft between the rows.

Keep the ground quite clean by frequent hoeing, and use great care when earthing up the rows. Lift the crop when the foliage starts to die down and the tubers are sufficiently set, to avoid rubbing off the skin.

As soon as the Potatoes are dug, select the most suitable and put them in a cool place with all light excluded. Do not wash the tubers until the day previous to the show, and carry out the operation either with a soft brush or a sponge, using cold water and plenty of soap.

Select only tubers which are uniform in shape and size. At all times, keep the tubers well covered after lifting until a few minutes before judging on the show-bench, to exclude light; failure to do so will encourage greening of the tuber, which detracts value from the perfect exhibition dish. Points value, 20.

Radish

Raphanus sativus

ANNUAL

Radishes have been eaten in this country since the fifteenth century and were grown in both China and Japan well before that.

It could be said with some truth that Radishes can be grown during every month of the year, especially if we consider the two distinct types which can be cultivated. For, in addition to the small-rooted spring and summer varieties there are some with large roots and strong foliage, known as Winter Radishes, which are usually sown in July for use during a large part of the winter.

SOILS   There are several things to avoid in the cultivation of first-class succulent Radishes, and the first of these is a poor impoverished soil. The ideal soil is one that is reasonably light, in good heart, but not recently dressed with rich manure, as this will produce a strong foliage at the expense of the roots. The land should be rich in humus from earlier dressings of manure or compost applied to the ground for a previous crop.

TIME OF SOWING   The first sowing outdoors is usually made in March, and a warm sunny plot should be selected, as the importance of quick growth cannot be stressed too strongly. Radishes grown slowly for various reasons can be most disappointing. Avoid a thick sowing on poorly-prepared land. Seed should always be sown thinly, about $\frac{1}{2}$ in. deep, and then covered with $\frac{1}{2}$ in. of fine soil, the surface of which should be firmed.

A March sowing runs the risk of severe frost, so it may be necessary to protect the young seedlings in such circumstances with a light covering of litter to a depth of 3 or 4 in.; or, of course, the use of cloches would not only protect the young seedlings but would result in quicker and steadier growth in the early stages.

Successive sowings can be made from March until September at intervals of two to three weeks, or according to the household requirements.

For the production of crisp, mild-flavoured Radishes quick growth is essential; therefore they must not be allowed at any time to become dry. Thin the seedlings early, as too thick a stand of plants will result in strong foliage and poor pithy roots, and added risk of early running to seed or bolting.

There is quite a wide choice of varieties, which vary in shape from round, oval and long. Actually Red Forcing is the earliest of all types. This is bright red in colour and round in shape, with very short foliage; ideal not only for open ground cultivation but also for cloches, frames, and greenhouses. Another popular variety is the Crimson French Breakfast which is half-red,

109

37. *Radish: Crimson French Breakfast, Red White-Tipped and Red Forcing. A selection of various shapes and colour, early maturing varieties.*

half-white in colour, white-fleshed, oval shape, and a few days later in maturing than the Red Forcing.

Many gardeners grow the Mixed Radish, which is a special preparation of varieites to mature at different dates, to extend the period of use and also to lessen the periods of successional sowing. Usually the names of the varieties are printed on the packet; the mixed colours and shapes can look most attractive in a salad.

Radishes are best sown in short rows in succession so that they can be kept weed free, which is not possible with a broadcast bed.

WINTER RADISH   The best month for sowing this is July, although August is usually not too late. Sow in rows 9–12 in. apart and thin the roots to 6–8 in. apart, as they are very much larger than the small summer Radishes and grow much more slowly. They can reach a weight of 1 lb each under good growing conditions. It is usual to leave them in the ground and pull as required, but in the more severe weather it is advisable to cover with straw or bracken. If necessary they can be lifted and stored in boxes of sand in a cool, airy place.

The roots should be sliced or shredded for use in winter salad, to which they are an excellent adjunct.

Winter Radishes may also be cooked in the same way as Turnips, and they are very tasty used in this way.

There is the choice of China Rose, a bright, rose-skinned, oval root with pure white flesh; Round Black Spanish, black-skinned but with white flesh; and Long Black Spanish, which is also black-skinned with white flesh.

The only insect pest likely to be encountered in most seasons with any type of Radish is the Flea Beetle, which can be kept at bay with Murphy's Sevin Dust.

Rhubarb

Rheum rhaponticum

PERENNIAL

Rhubarb was known in the time of the Romans and was cultivated in China as early as 2700 B.C., but it was used solely for medicinal purposes until about the eighteenth century. It was first grown in England in the sixteenth century.

SOILS   Although Rhubarb is said to be an accommodating plant, for first-class results it requires a rich deep soil, really well cultivated and, if possible, liberally dressed with manure, wood ashes, and bone meal. It prefers a sheltered position well away from trees.

CULTIVATION   The quickest way to make a plantation of Rhubarb is to purchase some roots of a first-class named variety and plant in one long row, 3–4 ft apart, or in a bed 4 ft apart each way, depending of course on the variety; some of the smaller kinds would require only $2\frac{1}{2}$ ft each way. Plant with the top bud 2 in. deep, tread fairly firm, and lightly prick the ground over. Planting in February is considered the best time.

The plants should be allowed a full season's growth before a single stick is pulled, and, even in the second season, pulling should be moderate, thus allowing the plants to build up a good strong crown.

Rhubarb is a crop that is very easy to force where there is a moderate warmth, as it is only necessary to pack the roots into boxes with peat, manure, and dead leaves, or any light soil. The roots will push into any moist material and find sufficient food. If exposed to the light, forced Rhubard has a full colour, but the quality is better and the colour good enough if it is forced in the dark. Therefore, if putting the roots under the staging of a greenhouse where there is a fair amount of light, it is advisable to place an empty box or barrel over it to produce some blanching,

or you can drape some sacking from the bench of the greenhouse to the floor, to exclude light.

A steady temperature of 45°–50°F is required until growth is steadily maintained, but the crop may be finished at 60°F. A higher temperature than this is not advisable.

Another quite straightforward method to obtain a slightly earlier crop outdoors is simply to cover the crowns where they grow; but this

38. *Rhubarb: Champagne. A popular market garden variety for the early pulling of dark red sticks of fine quality.*

111

operation will result in the roots so covered giving lighter yields in the next season.

There is one other method to produce a stock of Rhubarb, and that is by raising your own crowns from seed, but it should be pointed out at once that Rhubarb does not come true from seed and that the resulting plants will require a certain amount of 'rogueing' to remove some of the weaker growing specimens and to raise the standard of the stock. However, for those who are interested, seeds can be sown in the open ground during March or in a heated greenhouse in February. Prick out the seedlings when they have made four leaves, either into deep boxes or 4-in. pots or into a well-made bed in a frame or in the open ground under cloches. During early May, after removing the weaker-looking plants, plant out in the open for setting out the following spring into their permanent quarters; you will have to wait a year before beginning to pull.

On no account go on pulling sticks until there is nothing left. The pulling period should always be restricted to three months. Also, remove every flower stem as it appears.

It is well to remember that Rhubarb is a gross feeder. The bed must be dressed with a complete fertilizer in the early spring and, if possible, the crowns mulched with manure or compost after pulling has ceased. In addition to this, a dressing of a nitrogenous fertilizer should be given in May and July.

Salsify

Tragopogon porrifolius

BIENNIAL

Salsify (also known as the Oyster Plant) is one of the lesser-known root crops, although when well grown it can be of high quality. It can make quite a pleasant change from Carrot, Turnip, or Parsnip and the large roots are quite easy to prepare for cooking.

It is said to be a native of Algeria, and has probably been grown in this country for about 200 years.

Salsify has a cream-coloured root and slender, grass-like foliage. Fully grown specimens are usually 9 or 10 in. long, and 2 in. thick at the top. The roots have a taste very similar to that of oysters when prepared for the table; hence the name Oyster Plant.

SOILS   The best soil for this crop is a deep moist one with a rather light texture, although heavier soil will produce good roots, providing it is deeply cultivated and in a friable condition. Naturally, it should be fertile, but free from a recent application of manure in the top spit.

TIMES OF SOWING   Salsify should be sown in April or May. Sow the seeds one inch deep in rows 15 in. apart. Hoe frequently, but do not use a fork or spade near the crop, for loosening the soil can cause the roots to fang.

During November dig a portion of the crop and store in sand; lift further supplies as required. Take care not to damage the roots at lifting time as they will bleed and lose their flavour.

If you leave some roots in the land, these will produce chards in the spring. These are the flowering shoots which rise green and tender; they should be cut when no more that 5 or 6 in.

long and dressed and served in the same manner as Asparagus (*q.v.*).

Salsify roots need to be prepared for use by scraping them and then steeping in water containing a little lemon juice or vinegar. They should be boiled until tender and served with white sauce.

To prepare them as the 'Vegetable Oyster', the roots are first boiled and allowed to become cold. Cut them into slices and quickly fry in butter to a light golden brown, dusting with salt and white pepper while cooking.

39. Salsify and Scorzonera. These root vegetables, a delicacy for the table, crop well for use in late autumn and winter. See page 114 for Scorzonera.

Scorzonera

Scorzonera hispanica

PERENNIAL

This root is a native of Spain, and like Salsify, it has not been grown in England for more than 200 years.

SOIL Most soil well prepared will grow a useful crop of Scorzonera, but one of the best is undoubtedly a sandy soil with a coating of manure put in the bottom of a trench, taking care that there is none in the top spit.

SOWING Make a seed-bed and sow the seeds in April and May in shallow drills about 15 in. apart; thin gradually until the plants are a foot apart. Keep the hoe moving between the plants.

The crop should be ready for use in September. Lift as required, as you would with a Parsnip crop. To cook the roots, they should first be scalded, then scraped and put in water in which there are a few drops of lemon juice. Let them remain half an hour, boil in salted water in much the same way as Carrots until they are quite tender, and serve with white sauce. If left to get cold, they can be sliced and fried in butter to make a useful side dish.

Sea Kale

Crambe maritima

HARDY HERBACEOUS PERENNIAL

The *Crambe maritima* has grown in the wild state for many years in the sands and marshes close to the sea in the counties of Sussex and Dorset and in similar areas elsewhere. It is easily grown and is a crop which will give quick returns when well treated.

SOILS The soil for this crop should be exceptionally rich, and where possible enriched with good stable manure. The site should be prepared in the winter months, and deep digging is essential. The ideal soil is a sandy loam and, for those close to the sea, composted seaweed is an ideal addition; failing this, a first-class vegetable compost should give excellent results. As an alternative use shoddy with hoof and horn (at the rate of 2 oz per square yard). The land must not be short of lime, and, just prior to planting,

give a dressing of a complete fertilizer such as Growmore, at about 4 oz per square yard. Add plenty of sand to the heavy types of soil and, unless it is limey, give a dressing of carbonate of lime at the rate of 2–3 oz per square yard. Top dress with Sulphate of Ammonia ($\frac{1}{2}$ oz per square yard) three times during the growing season.

CULTIVATION Undoubtedly the quickest way to produce Sea Kale is to buy well-developed planting crowns, which are best planted during March. Set these at 24 in. apart each way and about 2 in. below the level of the soil.

However, the most economical way is to raise the plants from seed, for which you require the variety, Lily White. This method will take two years to produce forcing crowns. Sow the seeds 1 in. deep in an outdoor seed-bed in March or April in drills 12 in. apart. The seeds will germinate quickly and the plants will grow rapidly. During the following spring transplant the one-year-old plants to their final quarters. During the non-cropping years allow one bud only to develop per plant. Remove all flowering stems as soon as they appear. The growth of Sea Kale should be encouraged in every way; in the summer it will require water, liquid manure, and mulching with manure or compost to almost any extent.

A much quicker method is by the use of root cuttings, or 'thongs', as they are usually called. These are prepared from the clean straight side shoots which grow out from the main root. Choose the best of these side shoots—about the thickness of a pencil or a little thicker. Cut into pieces 6 in. long, the thicker end being cut level and the thinner end slanting. By this means the grower will know which is the top of the thong in the spring and can plant the right way up. After this preparation, the thongs should be tied up in bundles and placed in layers of damp sand. They can remain in this until planting time in March. When they are uncovered, the top end of the root should have made several eyes, and all of these except the two strongest should be rubbed off.

To utilize the ground fully, 'catch-crop' the bed with such subjects as Onion, Lettuce, Radish, or early Cabbage.

The bed of Sea Kale should be well looked after, feeding and watering as becomes

necessary, and, if available, give a mulching of strawy, rotted manure and good compost in early May.

As soon as the leaves wither, fork the ground over, and during November cover the plants with either large flower-pots or boxes and pack manure or decaying leaves over. This will bring along the blanched crop of Kale for eating. Another method is to surround the beds with boards and cover to a depth of 1 ft with leaves—the boards will prevent the leaves from blowing away.

Sea Kale is ready for cutting when the blanched growth is 7–8 in. tall.

FORCING SEA KALE   This is quite a simple matter. You must have perfect darkness, and the maximum temperature for this is 60°F at any time. You can begin at 45°F, which can rise to 55°F to ensure a satisfactory growth.

Sea Kale slowly forced can be almost as good as that grown under pots in the open without any heat at all. Spare pits or odd places which you have available can be used for this crop, providing the heat is not too great. Pack the roots in mould or leaves, or if you have it, half-rotted manure, and then shut them up to exclude light. The crop should be ready in five or six weeks.

Once roots have been lifted for forcing they should be thrown away when the crop has been cut, but you will find roots which have been forced in the open ground suffer very little, so that they may be used for several years before it is necessary to renew them.

Shallot

Allium ascalonicum

PERENNIAL

Originally a native of Palestine and first introduced into England in 1548, this is a hardy bulbous plant which is used as a vegetable in this country.

SOIL   The Shallot is not a difficult crop to grow and it succeeds best in an open position and in a lightish soil, deeply cultivated and well drained. For satisfactory results the preparation and manuring should equal that for the Onion (*q.v.*) when the latter is grown only for the kitchen. Too rich a soil can result in producing Shallots somewhat too large for pickling.

PLANTING   The old-fashioned custom of planting on the shortest day is not often practised these days, for most gardeners will find early spring is the best time for general purposes, and it is usually February before a start can be made. The soil should be in a friable condition and usually should be trodden firmly, as is recommended on most soils for Onions.

A recommended planting distance is 4 or 5 in. between the bulbs, with the rows 12 in. apart.

Press the bulbs into the soil and then firm around them as the growth of the roots tends to push them out of the soil. It will be advisable to examine the bed from time to time to see what root disturbance has taken place, and then put the matter right.

Hoe the crop from time to time, but do not bury the bulbs in so doing as Shallots grow best on the surface of the soil.

Ripening should commence during July when the leaves commence to turn yellow, and, as soon as they have withered considerably, the bulbs should be lifted. Spread them out to dry on the soil, turning them from time to time. They can then be divided, topped, and stored in a dry cool place.

VARIETIES   The best type for ordinary garden cultivation for the kitchen is the Giant Yellow, which is hard-fleshed and a fine keeper. However, if the variety is required for the show-bench, it will be necessary to obtain a good stock of Hative de Niort, which possesses perfectly symmetrical shape and a marvellous ripened skin-colour. R.H.S. points value, 12.

Spinach

Spinacia oleracea

ANNUAL

This plant is said to be of Persian origin and it was originally grown solely for its medicinal properties.

SOIL   Spinach will grow on most soils but it does require well-prepared land in a good state of fertility, land which has been manured for a previous crop, so that the plants are constantly growing, without the slightest check in growth, as this so often results in bolting or premature

40. Spinach: Greenmarket. This unique variety can be sown at all periods of the yar. Heavy foliage, slow to bolt.

running to seed, which in turn means a complete crop failure. The soil should be well supplied with potash. Spinach will not succeed on very acid soils, and a *pH* check should therefore be made and, where necessary, lime applied and well mixed with the top soil.

What is required is quick growth and a few heavy pickings of large succulent leaves, and this is brought about by skilful cultivation.

SUMMER SPINACH In favourable weather and soil conditions, the first sowing can be made towards the end of February or early March, especially on warm soil in a sheltered sunny spot. Further sowings can be made in succession, at intervals of two to three weeks, until the beginning of July. Small and frequent sowings are to be recommended because of the possibility of Summer Spinach running to seed very quickly, especially in hot dry weather. Try to provide rich moist soil conditions, and, most important, allow plenty of room. Thin the seedlings as soon as large enough to handle, leaving them 3 in. apart when quite small; make another thinning later to 6 in. apart; and finally leave at 12 in. plant to plant. The drills should be 12–15 in. apart. In this way, each plant will cover the space, which should produce fine large leaved plants. When pickling, take two or three of the largest outside leaves from each plant.

TIMES OF SOWING To be able to provide a succession of pickings from October to May (when the early spring-sown crop should be ready), three sowings should be made from the first week of August to about the middle of September. The site for Winter Spinach should be a well drained one. Sow in drills 12–15 in. apart, 1 in. deep, and thin the seedlings as quickly as possible, leaving in the first instance at 3 in. apart and later 6 in., and then leave at this distance throughout the winter. The August sowing would benefit by gathering the first leaves, and then leaving the plants throughout the winter to produce an early flush of foliage.

Protection by the use of cloches is a wonderful advantage during the worst of the winter weather.

It is advisable with Summer and Winter Spinach to gather the crop in such a way that you never strip the plants but leave as much foliage to remain as you pick off. Take the largest fully-grown leaves, but do not let them get old and tough by leaving them on the plant too long.

VARIETIES For Summer Spinach select the variety, Sigmaleaf; for winter, Greenmarket is an outstanding type.

EXHIBITION   Judges often throw out an exhibit because the exhibitor has not read his schedule carefully, staging Perpetual Spinach when the wording of the class definitely calls for Annual Spinach.

For an exhibit of Spinach in the competitive classes, not more than 25 leaves are staged, either in the single dish class or in a collection.

Spinach, Perpetual, or Spinach Beet

Beta vulgaris cicla

BIENNIAL

This is a hardy biennial plant which has a similar flavour to the annual forms of Spinach, both the round and the prickly-seeded. Perpetual Spinach is usually grown for autumn and winter gathering, and the leaves are much larger and somewhat more fleshy than those of the ordinary Spinach.

Spinach Beet is, in fact, one of the leaf Beets which produces an abundance of edible leaves and has a fibrous root, unlike the usual types of Beet, and this root has no culinary value. The plants are moderately hardy and can bear a few degrees of frost and so will continue to crop until late in the season.

TIMES OF SOWING   This may be carried out from April until the end of July in rows 15 in. apart; thin the young plants to a distance of 8 in. in the rows. When the large outer leaves are ready for gathering they should be removed, whether wanted or not, so that a strong, continuous growth is promoted.

SOIL   A really rich, well-drained soil is required to produce a good-sized, strong plant capable of cropping over a very long period; irrigation, or watering, is a great advantage in very dry summer periods.

It has been found that Spinach Beet is

41. *Spinach: Perpetual. Often referred to as 'Cut and Come Again'; gives a continuous supply from July until the following spring.*

particularly useful on dry soil where Annual Spinach runs to seed prematurely.

Silver or Sea Kale (Swiss Chard) is another of the leaf Beets, with no edible root but with large, pale green leaves which are used as Spinach, of first-class flavour. In addition, this plant has broad, white mid-ribs which are delicious when cooked and served in a similar manner to Sea Kale (*q.v.*). It is grown in precisely the same way as Spinach Beet.

One other interesting and useful Leaf Beet is the RHUBARB BEET which has red leaf stalks and mid-ribs and is frequently used for decorative purposes in flower borders.

Spinach, New Zealand

Tetragonia expansa

ANNUAL

New Zealand Spinach is a native of Australasia and is said to have been introduced into this country in about 1772. Most people agree that this type lacks the bitterness of the true annual Spinach; and it has one other advantage: that it will grow in the driest and hottest summers when ordinary Spinach usually bolts to seed very quickly, especially on the poorer soils.

SOIL   The best type of soil for New Zealand Spinach is a medium light one in a good state of fertility.

TIME OF SOWING   Seeds may be sown outdoors at the beginning of May or under glass about the end of March. If possible the plants should be potted singly and, after hardening off, planted out towards the end of May. Planting distance should be at least 3 ft apart each way in a sunny position. The plants will ramble some distance and soon cover a great deal of space. The growth is rapid and the plants must not be kept short of water. In about five or six weeks the first lot of succulent leaves should be ready for gathering. The quick growth will continue throughout the summer, and, apart from watering, no attention is required.

The leaves are best picked singly and should be stripped off the stem. This will allow the shoots to continue growing, thus giving a continuity of supply.

Strawberry

Fragaria

HARDY AND HALF HARDY PERENNIAL

The Strawberry is a fruit, yet this plant is grown in most kitchen-gardens, and the general routine of work has to be arranged with this crop in mind.

The Strawberry is undoubtedly the most certain of all our hardy fruits and is much appreciated for eating fresh in the summer as a great treat and also a preserve for use in the winter. Most people would agree that it deserves the best of cultivation; its demands are few and sometimes, under the poorest system of management, it can be very prolific.

There is the choice of seeds, divisions and runners for making a plantation of strawberries, but undoubtedly the best way is by planting rooted runners of reliable, named, virus-free varieties, in an open, sunny spot, in really well-prepared ground during the spring or autumn, when fresh and good runners are available. It will be found that late planting is not worthwhile, for, if the plants have insufficient time to establish themselves before winter sets in, many will be lost. If you plant during the spring and remove the flower-stems, they should yield a heavy crop in the following season.

SOIL TREATMENT Strawberries require a deep, rich soil, although good crops can be grown in soils of varying character if thoroughly well prepared. This crop may occupy the same plot for some considerable length of time, so it is essential that the soil should be made thoroughly fertile in advance of planting by deep digging and the use of plenty of manure or compost.

Where available four barrowloads of farmyard or stable manure per square pole is by no means excessive for garden cultivation. If your soil is very poor or is light in character, an even larger quantity can be used, for natural manure greatly assists in retaining moisture in the soil during a dry period. Strawberries will not thrive in a dry soil.

In addition to the manure or compost do not hesitate to add a first-class complete fertilizer raked into the surface of the soil at planting time, as this is a practice largely adopted by commercial growers.

Once the plants have obtained a firm hold on the soil it does not matter how hard the soil becomes. It is one of the secrets of successful Strawberry cultivation that the bed should be firm and compact, and, when forcing plants, the soil should be positively rammed into the pots. When a new plantation is made, make sure that it has good drainage, for, whilst the Strawberry thrives best in a moist soil, yet stagnant water is fatal to its well-being.

If the Strawberry plants you have obtained are rather dry, unpack them at once and spread them out thinly in a cool, shady spot and sprinkle lightly with water to revive them.

When planting, take care to spread out the roots so that the plants have every opportunity to develop properly into strong, healthy-looking specimens. Do not bury the plants too deeply and be quite sure the crown is just clear of the surface level.

DISTANCES TO PLANT Most growers allow 2 ft between the rows and 1½ ft between the plants in the rows, but much depends on the character and strength of the variety.

Sometimes the plants are put a foot apart each way and, after the first crop of fruit, every alternate row is removed, and then every alternate plant in each row is taken out. This

120

leaves the remainder at 2 ft each way. The soil is then lightly forked and a good coating of manure put on.

During growth keep down the weeds, supply plenty of water in dry weather, especially when the fruits are swelling, and always remove the runners as fast as they appear.

Apply a good coat of long, strawy manure in February. In doing this, no harm will be done if the plants are partially covered, as they will soon push up and show themselves; and by the time the fruit appears the straw will be washed clean or fresh straw can be used later. The crop thus helped should be a heavy one.

As soon as the fruit shows some sign of ripening cover with netting against the attacks of birds.

When the whole crop is picked, remove the old large leaves; this is best done with a knife. This operation will admit air to the young leaves, and on the free growth of these the formation of good crowns for the next year's use depends. In this way root action is promoted and the embryo buds are formed that will in the next summer develop into Strawberries.

If there is no natural manure available for top dressing, artificials should be raked in around the plants as soon as growth commences in the spring.

Many gardeners destroy their Strawberry-bed every three years, but a better idea is to make a small plantation each year and at the same time dig up the old plantation that has served its turn. In other words, add a row or two on one side of an existing bed and do away with an equal number of rows on the opposite side. A change of stock from time to time can be very desirable, for the plants deteriorate when grown for many successive seasons.

Some years ago a number of growers raised their own Strawberry plants from seeds sown under glass very early in the year, using that old favourite, Royal Sovereign, but such a practice is no longer adopted, as all growers purchase rooted runners from a reliable supplier, to ensure commencing with virus-free stock.

Alpine Strawberries are very largely grown in France, probably more so than the large-fruited varieties which are so popular in this country. The best method is to sow the seeds in January in pans filled with a light, rich compost, and place in a gentle heat. Prick out the plants on to a bed of light soil in a frame, or into seed-trays. As soon as the plants are large enough, usually about the end of May, they should be transplanted into the open ground. From these sowings fine fruits may usually be gathered in the following September. When a full crop has been gathered the plants should be destroyed, a succession being kept by sowing annually. By slowly growing the plants from spring-sown seeds and potting in the autumn, it is not a difficult matter to have this type of Strawberry in fruit under glass at Christmas.

The above treatment of culture can be used for the growing of the true wild Strawberry known as Fraise des Bois.

Swede, Garden

Brassica rutabaga

BIENNIAL

Swedes are extremely hardy in constitution and the roots will keep several months longer than any variety of garden Turnip; they also have the advantage of being extremely easy to store for late winter and spring use.

The culture of garden Swede is in all respects the same as for garden Turnip (*q.v.*), and the young seedlings are equally as attractive to the Flea Beetle, which means that a careful watch should be kept for this pest from the time of germination until the plants have become well established. An immediate application of Sevin Dust is necessary as a preventative or as soon as this pest is noticed.

The date of sowing depends on the district in which they are to be grown. In the north of England early May is usually the best time for this operation, whilst in the Midlands and the southern counties late May or early June is recommended. In districts where this subject

42. Swede: Purple-Top Garden. Almost neckless, hardy, yellow-fleshed roots.

suffers from bad attacks of Mildew in the foliage, useful roots for winter use can be obtained from a July sowing; it should be remembered, however, that from this late sowing one cannot expect large roots. However, a healthy, smaller-sized root does give better table-quality, and Swede sown at this period of the year usually produces roots of very fine flavour.

Sow the seeds in drills about $\frac{3}{4}$ in. deep, 15 in. apart and thin by degrees to some 12 in. plant to plant.

It is most important in dry weather periods to give the rows of Swede a really good soak from time to time so that continuous growth may be maintained. This is a crop that, once it has received a severe check in growth, may be most disappointing in its ultimate results.

The use of the Dutch hoe from time to time may also assist in maintaining steady growth. Under average growing conditions Swedes will require fully 25 weeks to produce sizeable roots for culinary use.

CHOICE OF VARIETIES For those growers in districts troubled with Club Root and Mildew, the ideal type is Chignecto, which has been specially bred for resistance to these troubles, in addition it has a particularly fine neck with a pretty, round-shaped root, with deep purple skin colour.

For areas free from disease, the variety Purple-Top has been carefully selected over the years for its almost neckless character and deep yellow-fleshed roots and pleasant, bright purple skin. The very latest to mature is Bronze-Top, extremely hardy, with solid yellow flesh; it develops very slowly indeed. This variety has a purplish bronze skin.

HARVESTING AND STORING Garden Swedes can be left in the ground without suffering winter damage much longer than any other root crop, being so hardy in character; but for those who wish to store, the first operation is to cut off the tops, but not too close, thereby leaving a little of the green neck, and store in a like manner to Turnips.

Sweet Corn

Zea mays

ANNUAL

Sweet Corn, Sugar Corn, or Corn-on-the-Cob is one of our sun-loving plants and the best crops are usually grown in the southern half of England. It is a curious plant as it has to produce a certain number of leaves before it will flower. The male flowers appear in spikes at the top of the plant and the female flowers grow lower down. The varieties used in this country are dwarf in character, having few leaves to make before they flower, and these mature very quickly.

SOILS This crop should be given a deep soil in a sunny, sheltered position. Medium to light land is considered the best for Sweet Corn. Fresh manuring is not recommended because it can produce an excess of nitrogen which induces too much leafy growth.

TIMES OF SOWING Seeds may be sown *in situ* by about mid-May in the south; sowing before this may result in crops not doing well, and they may easily suffer from late May frosts. For

43. Sweet corn: First of All. One of the dwarfest growing varieties, early to mature.

growers with cloches seeds can be sown from about mid-April onwards.

Sweet Corn may also be sown in pots or soil blocks in a cold frame or greenhouse around May 1st, and these should be ready to plant out at the end of May or early June. Seed-trays are not suitable for the raising of Sweet Corn. The plants require a distance at least 15 in. plant to plant in the rows, and not less than 3 ft between the rows.

Birds, especially rooks, can be a serious trouble, for they will either remove the seeds or sometimes even pull up the young plants. It is advisable to protect the plants with black cotton stretched across the rows.

One other possible trouble is an attack by the Frit Fly, an extremely small insect which lays its eggs on the earth around the young seedlings soon after they appear through the soil. Larvae will hatch from the eggs and feed on the seedlings, near the growing point, so that the plants will become stunted and distorted. These plants should be removed at once, as they will never produce a worthwhile crop.

The crop benefits from a regular hoeing, and ample supplies of water during dry periods are necessary.

The cobs become ready for picking about one month after the silky tassels show and just as these begin to turn brown. The corn is ready for eating when the seed resembles a creamy consistency, which can be tested by squeezing between the thumbnails, unfolding the outside wrapper or leaf around the grain.

The cobs can either be cut from the stem or snapped off with a downward jerk. An average crop is some three to four cobs per plant.

Sugar Corn requires using soon after picking as the seeds will rapidly dry out, and it is imperative to keep in a cool place prior to cooking.

CHOICE OF VARIETY   The hybrid strains are to be preferred to straight varieties, such as Golden Bantam; and First of All, which matures very quickly, is by far the best for the weather conditions so frequently experienced in this country; and Kelvedon Glory makes a good succession to this.

When Sugar Corn is intended for exhibition, select fresh, cylindrical cobs, well set with grain in the best possible condition for table use and of a uniform colour, which may be yellow, white, or some other colour, according to variety. The maximum number of points awarded is 15. Six cobs are required in a collection and three cobs for a single dish.

Tomato

Lycopersicum esculentum
ANNUAL

Although the Tomato has been known in this country since the sixteenth century, the shape and type of fruit of varieties up to the commencement of the present century left a great deal to be desired. The old varieties were considerably larger, flatter in shape, and completely corrugated. Nevertheless, they were fleshy and more in keeping with some of the varieties still grown in continental countries and the Americas.

During the last seventy to eighty years it has become one of the most popular vegetables throughout this country, whereas before there was little demand for it from the consuming public. Indeed, it would be true to say that the whole character of the Tomato as we know it today has been developed in this present century. Today most varieties are smooth, round, varying on average from six to eight to the pound in size. But it is doubtful if there is one single vegetable which can compare with it in variations of methods of cultivation—heated greenhouses; cold greenhouses; polythene structures; in the open ground outdoors; growing in pots and boxes under walls; on verandahs; ring culture; growing in Tom or Gro Bags; or growing on straw bales. Besides this, when you consider all the troubles and problems which beset the grower—diseases, etc.; feeding; watering; ventilation; the choice of soil or peat compost—few would disagree that, in producing a crop of Tomatoes, you have met a challenge, and have also brought off a tremendous achievement. Having said all this, the object of this chapter is to assist in producing a first-class crop!

Here is a case where it would be remiss on any author's part not to acknowledge the splendid work accomplished by research workers on breeding in recent years. It is due to their untiring efforts primarily in the interests of the many commercial growers, both here and in the Channel Islands, that have found the answers to many of the problems experienced years ago in Tomato-growing have been found to the great benefit of the amateur grower of today.

SEED-SOWING AND PLANT-RAISING   It will be appreciated that, with the many available methods of producing a crop, dates of sowing can vary from January to May. Obviously, the earliest sowings are for the crop that can be given heat, and the latest for outdoor cultivation. Nevertheless, the principle of sowing and raising is the same. A simple approach is to fill a flower-pot or a seed-tray to within about $\frac{1}{2}$ in. of the top with John Innes Compost No. 1, which should be pressed down and well watered before sowing. Immediately the seed is sown it should be covered with $\frac{1}{4}$ in. layer of the same compost. The whole should then be covered with a sheet of glass and a layer of paper to exclude the light during germination, removing the covering immediately the seedlings appear. The temperature should be approximately 65°F whilst waiting for the seedlings to appear, and then lowered to 55°F or 60°F. High temperatures with early sowings of some varieties can be dangerous, as this would encourage 'Jack' plants, which will be referred to later (see p. 128). Immediately germination has taken place the pots or boxes should be placed in a position where maximum light is available and as close to the glass as possible, thus ensuring sturdy growth of the seedlings. A watering with a fine rose preferably early in the morning is a good thing.

When the seedlings are about $1\frac{1}{2}$ in. high, they should be pricked out into 3 in. pots again, using John Innes Potting Compost No. 1, making sure that the compost is thoroughly moist. After pricking off, place the pots on the staging of the greenhouse in good light. No further watering is necessary for two to three days, but an overhead spraying may be desirable. Maintain a temperature of 55°F to 60°F, making sure to ventilate the house whenever possible. It usually takes about a month from the time of pricking off before the plants are ready for transplanting to their final quarters.

If transplanting in the border of the greenhouse or under polythene structures, the plants should be spaced 15 in. apart in the rows with 20 in. or a little more if possible between the rows. If transplanting into pots an 8-in. pot is enough, but a 12-in. pot is better. When growing in pots or boxes the safest method is to use a John Innes No. 3. Whichever method is used, it is timely at this stage to remind you that, before transplanting, you should give the pots in which the plants are growing a good soaking to make sure they are not dry at the ball, as, after transplanting, it is always difficult to get the water to the roots of the plant. This applies equally if transplanting outdoors.

SOILS   Many disappointments experienced by the amateur grower can be traced back to neglect in several directions, but in none more than in choice of soil. If you are in doubt as to the health of the soil in the greenhouse, or of that which is brought in, it is a wise move to sterilize it. Further, if one is doubtful on the question of its physical condition you cannot do better when transplanting, having opened up with the trowel a hole of sufficient size to take the plant, than to dust this with a base Tomato fertilizer, to ensure a good sturdy growth. The plants should be watered uniformly at all times, but further feeding must wait until the bottom truss of fruit is set. Here again it is very important to remember, when climatic conditions are favourable, to give as much ventilation as possible.

The method of support of the plants depends on how the crop is to be grown. For example, if planted in the bed of the greenhouse, the best arrangement is wires suspended from the roof, with strings tied to the wire and brought down to the plant and tied to a short stake. On the other hand, if grown in pots or boxes a good stout bamboo will serve the purpose, the length depending upon how many trusses you are prepared to grow, whilst, with an outdoor crop, a 4–5 ft cane is long enough, for the simple reason that, in English climatic conditions, it is advisable to stop the plant at the fourth truss. After the first truss is set, regular feeding at twelve or fourteen day intervals is essential.

Another method of cultivation used by many growers is ring culture. The principle is to grow plants in a relatively small amount of soil, in bottomless pots, standing on an inert, well-drained base. Feeding roots develop in the soil, whilst a vigorous secondary root-system grows out through the base material, which must be kept well watered; thus the plants have access to ample moisture without the danger of root-rot, which can occur in normal soil culture. Well-weathered cinders, varying from finely crushed to about $\frac{1}{2}$ in. in diameter, has been found to provide the best base medium. A layer of between 5 and 6 in. is recommended, and this should be placed on a firm foundation, making certain it is well drained, but at the same time retains sufficient moisture to supply the plant's requirements. Alternative materials which can be used are sand, $\frac{1}{4}$-in. gravel, peat, and vermiculite, but of these coarse sand and gravel are the best. Fine sand is not suitable as it will pan, thus excluding air.

Nine-inch bottomless pots are the ideal size to set out on the prepared base, 15–18 in. apart. It is important to remember that these should be filled *in situ*, firming the soil thoroughly to ensure a close contact between the compost and the base material. A medium compost such as John Innes No. 3 is ideal for the purpose.

When the plants are set in their pots, they should be well watered in, but subsequently little watering should be needed. However, the base material must be thoroughly soaked every day and, provided there is a close contact between pots and base, this should keep the soil moist. It should be remembered that, once the roots have penetrated to the base, the pots must never be moved or the root-system will be severly damaged.

Feeding should not be commenced until the first truss of fruit has set; thereafter, apply a weekly feed of about $2\frac{1}{2}$ pints of diluted fertilizer for each 9-in. pot.

GROWING ON TREATED STRAW BALES This method of culture is still very much in the experimental stage. The general procedure is as follows:

Bales should be placed along the greenhouse borders where the beds are to run. Sheets of polythene may be laid down first if it is desired to prevent rooting into the underlying soil. The bales are best laid on the flat as they are more stable than when placed on edge. Sufficient water must next be applied to wet the straw thoroughly throughout the bales, and this can most effectively be done by several applications spread over two or three days. On an average, about 9 gallons are needed per bale. This preliminary treatment should be commenced about eighteen days before the intended planting date, and all ventilators should be closed. A temperature of 50°F at least is necessary to ensure fermentation.

Two or three applications of a nitrogenous fertilizer are needed to induce fermentation, with intervals of about five days between, and Nitro Chalk is a good material to use, allowing $1\frac{1}{2}$ lb per bale (based on an average bale of 56 lb). The fertilizer should be sprinkled over the surface and lightly watered in. With the second or third application, 1 lb each of Sulphate of Potash and Superphosphate should be included, together with 4 oz of Magnesium Sulphate (Epsom's Salt), and 3 oz Iron Sulphate.

With adequate house temperature, fermen-tation will commence rapidly, with the centre of the bales heating up to 110°–130°F within three to five days. Soon after the last fertilizer dressing, temperature should commence falling and preparation for planting can be made when it is below 100°F.

Bales are generally arranged in lines along the length or across the width of the house, in a manner that permits a double row of plants to be set on each line of bales, at the normal spacing. Small beds are made up from sterilized soil from old cucumber beds, John Innes Potting Compost, or other suitable material, in which the plants are set.

As the bales rot a certain amount of sinking will take place, so plants should not be too firmly tied to supports or the roots may be pulled out as the substratum sinks.

Attention to watering is of particular importance as bales may dry out quickly. Watering every day may be necessary in bright weather. The usual feeding with a Tomato Top Dressing or Liquid Feed should be given as growth proceeds.

VARIETIES Many amateur growers must be really confused when confronted by a seedsman's catalogue which includes a long list, making the final selection most difficult. Therefore, a word of advice at this juncture may prove of assistance in making the final choice. Apart from the bush dwarf-growing varieties and the ornamentals, Tomatoes can be divided into three groups:

1 Varieties suitable for early sowing and growing on in heat.
2 Varieties for growing under glass or polythene tunnels, cold.
3 Varieties which can be grown either in a cold house or outdoors, where quality, flavour, and texture take precedence over quantity.

To fill the requirements in the first group Eurocross BB is outstanding, but many of the other F_1 Hybrids will produce first-class crops.

44. Tomato: Alicante. Suitable for all methods of cultivation: greenhouse, pots, boxes, ring culture, borders and outdoors. Heavy cropper, smooth round fruits.

In the second group, varieties such as Alicante and Moneymaker will fill the bill, but in the third group, where quality is the keynote, such varieties as Early Market, Ailsa Craig and Harbinger would be the choice. For outdoor cultivation with quality, Leader is outstanding, but, if quantity is preferred, Alicante and Moneymaker will again be a good choice.

BUSH TOMATOES Many amateur growers prefer the bush type, generally smaller-fruited but often quicker in ripening; and they are only really suited to outdoor cultivation. A simple way of producing a crop is by planting in the open ground, placing a layer of straw to keep the fruits clean, and allowing the plants to grow perfectly naturally without removing the laterals.

JACKS OR FEATHERHEADS Earlier in this chapter we referred to the 'Jack' plant, which can best be described as one of the freaks of nature. It is a plant which produces, right from the pricking out stage, feathery foliage. Before it is 4 or 5 in. high it shows an abundance of spikey lateral growth at each leaf axil. Experience shows that this can be found in some varieties more than others, when sown say in January or February in very high temperatures **and** shorter hours of daylight. If these plants are **grown** on, they are completely useless, producing just small sterile fruits; so watch the temperature in the early stages very carefully to avoid this trouble.

GRAFTING   In recent years, particularly in intensive commercial production areas, such as Holland and elsewhere, where Corky Root, Nematodes (Rootknot Eelworm), Verticillium Wilt, and Fusarium are troublesome, grafting on to root stock has been the answer, but it is the author's considered opinion that, if the amateur grower maintains a healthy soil and strict hygiene, it is unnecessary for him to undertake this difficult and laborious operation.

EXHIBITION FRUITS   The ideal dish for this purpose, whether in the single dish class or in a collection, should be selected with these factors in mind: uniformity of size; a fresh green calyx left on the fruit; freedom from all blemishes, such as Greenback and Blotchy Ripening; and the fruits should be firm in texture, the test being that the fruit does not give when pressed with the fingers—overripe fruits should be avoided at all costs.

DISEASES, ETC.   The Tomato plant is one of the most sensitive of all vegetables. Many troubles, failures, and disappointments can be traced back partly to human failure in understanding the essential plant requirements and partly to carelessness in general cultivation. The following list covers most common troubles:

Seedlings Damping Off   Experienced early in the season, this is due to sodden wet conditions, and can be avoided by using a compost which has plenty of humus, with clean sand to keep it porous.

Root Rot   Extreme wetness with poor drainage, together with coldness of the soil, is responsible for this trouble. On examination of the root, you will find brown decayed portions, which, as one would expect, soon affect the condition of the plant.

Wilt   During the heat of the day some plants will quickly flag, but partly recover in the cool of the night. This is due to insufficient attention at the time of preparing the soil. A good healthy compost, with plenty of humus, would prevent this trouble.

Leaf Mould (Cladosporium)   This disease only attacks crops grown under glass. It commences on the upper side of the leaf as a grey mould; eventually the leaves shrivel up. It is usually found in a greenhouse with an extreme moist atmosphere. This is due entirely to lack of ventilation and lack of through circulation of air. The lesson to be learnt here, is not to shut the house up day and night, but to give the plants all the air permitted by climatic conditions.

Grey Mould   This is a fungus disease which can be troublesome in the greenhouse. Is usually found at the base of a badly-pruned leaf-stalk, and again is due to extensive moisture and poor ventilation. It affects not only the foliage, but also the fruit; if a plant is discovered to be infected in the early stages, pull it up and burn it to prevent the spread.

Leaf Scorch   Another common trouble experienced by the amateur grower, it can be caused by using sprays which are too strong or fumigating a house incorrectly. Another form of Scorch is due to poor root-action induced by poor soil conditions, due to insufficient water being taken up by the plant on extremely hot days. If caused by sprays or fumigation it can be identified on the edges of the leaf; whereas Scorch through water shortage, the whole leaf can show it.

Dry Set This is the condition where trusses of flower have set, but, when the fruits reach the size of a fingernail, they fail to swell. It is due to poor pollination because of a very dry atmosphere.

Blossom-end Rot One of the most common troubles experienced by the amateur grower, this is easily identified by the large black portion at the flower end of the fruit. This is caused by a serious water shortage at a critical stage in the development of the young fruit. Immediately this is seen on a plant, remove the offending fruits, increase the water and feed, and you will find that later trusses will be free from this trouble.

Greenback In some varieties this trouble is often seen and is identified by the hard, discoloured tissues on the shoulder of the fruit. In this case there are two contributory factors: (1) a shortage of Potash, and (2) the exposure of the fruits to the direct rays of the sun. Shading of the house and spraying the glass of the house will prevent this trouble.

Cracking of the Fruits When the fruit cracks in the later stages of development, it shows in a circular fashion around the shoulder of the fruit. The cause of this is that, when the small fruits were developing, the plants were allowed to become dry at the roots; the small fruits harden. Plenty of water and nitrogen are then given, and these cause quick growth of the fruit; hence the splitting. This may well happen when the amateur grower has gone off for the weekend and taken a chance that all would be well.

Blight This only affects outdoor crops; the Tomato and the Potato are brother and sister. When humid conditions exist, this disease can sweep through the Potatoes and on to the outdoor Tomatoes; but it can be easily prevented by spraying the entire plants with Bordeaux Mixture, which is perfectly harmless. Use this as a preventive; do not wait for an attack in the hope of stopping it.

Finally, we must refer to hormone infections, which can completely distort a crop. The soundest advice which can be offered is, if you are spraying a lawn or weeds in the path, make sure that all doors and vents in the greenhouse are completely closed whilst carrying out this work, to prevent any possible wind drift affecting the crop. The Tomato is almost like a magnet to infection and is more readily affected than any other plant. Make sure also that any utensils used for hormone weed-killers are never used for watering in greenhouses. Last, but not least, any grass cuttings or weeds which have been treated must never be placed in the compost-heap.

The four important factors to remember in successful Tomato cultivation are soil, water, feeding, and ventilation.

Turnip, Garden

Brassica rapa
BIENNIAL

This subject is a hardy biennial of European origin, which is grown for its fleshy roots and also for its tender growing tops which are generally consumed in the winter and spring. It has been grown in England since about the middle of the sixteenth century.

SOILS   The ideal type of soil for the production of the best Turnips is a deep, sandy loam, which has been maintained in a good level of fertility. As with other types of roots, such as Beet and Carrot, fresh farmyard manure should not be used in the preparation of the plot required for Turnips. In common with other fast-growing plants of the cruciferous order, Turnips must have lime in some form, so in many gardens a good dressing of lime will be necessary. Both Superphosphate and Bone Meal are valuable in the soil preparation for this crop. It will be found that the majority of soils which are not too sour and not deficient in phosphates will produce quite good crops.

TIME OF SOWING   In most seasons a sowing can usually be made in the open ground, especially in a sheltered spot, by the end of March or early April, but it may be necessary during the worst of the weather to give some form of protection to the young seedlings, in this respect, cloches are particularly useful.

Naturally, sowings made during April and May will be less hazardous, as by this time the weather should have improved and the soil become warm, with fewer and less severe frosts.

The most important month for sowing this crop is usually July. Such a sowing should provide a supply for the autumn and early winter months. There is no reason why a later sowing cannot be made in early August, which

should ensure a most useful crop in the depth of winter.

Many gardeners have found that land cleared of Broad Beans and Peas can be sown with Turnips with very little preparation, apart from lightly breaking the surface and cleaning with a Dutch hoe, and following this with a light raking

45. *Seven different types of garden turnip: 1. Greentop White, 2. Golden Ball, 3. Sprinter, 4. Snowball, 5. White Milan, 6. Red Globe, 7. Jersey Navet.*

CULTIVATION Draw the drills about ¾ in. deep and about 12 in. apart; sow the seeds not too thickly and lightly cover with finely pulverized soil. It will be found that germination takes place very quickly if the soil is sufficiently moist. It is possible that the Flea Beetle will soon cause trouble in the seedling leaf stage, almost as soon as the tiny seedlings break the surface of the soil; it is imperative to apply a dusting of a good Flea Beetle Dust such as Sevin at the earliest possible moment, and to repeat the operation from time to time as may be necessary, according to weather conditions.

Try to thin the small plants early and leave them at a distance of some 6–9 in. apart in the rows.

It will be found that Turnips are most disappointing when grown under dry conditions, and water must therefore be given during dry periods when there is little prospect of rain. This of course is of the greatest importance with crops grown on light well-drained soil. When watering, do not just moisten the surface of the soil but give sufficient to go down to the roots, as frequent watering is not advisable, for it encourages strong foliage at the expense of the edible root.

It is well to remember that continued dry conditions for Turnips will result in roots which are misshapen and of poor quality and very inferior flavour.

SELECTION OF VARIETIES For the very earliest roots, choose either White Milan or its purple-topped counterpart, Sprinter, both of which are ideal for March and April sowing, either in frames or in the open ground. To follow these, try Snowball, which is round in shape, with pure white flesh and white skin; it is a little larger than Milan.

For the latest cropping and for winter storage, there is the well-tried Golden Ball, which has a much harder flesh than the Milan or Snowball types and has round, deep yellow roots; it can be left in the soil longer than most other varieties. It is outstanding for flavour and as a keeping variety.

Lifting and storing should be carried out in the late autumn or early winter. To carry out this operation really successfully the foliage should be cut off, but not too close, and the roots shortened, but not removed altogether. Roots are best lifted in dry weather and usually not later than the end of October.

Store in boxes or bins in sand or peat or on the floor of a shed. Place a 2-in. layer of sand or peat and follow this with a layer of Turnips, then a layer of storage material, and so on. Very few varieties of Turnips are really hardy and they do not equal the Swede in this respect; storage therefore should take place before the usual winter weather conditions commence.

There is no doubt that the best-flavoured Turnips are those which have been grown quickly; with the early varieties a period of eight to ten weeks is sufficient under ideal conditions to produce some fine young succulent roots.

Turnips grown for their tops or foliage alone are usually sown in July and August and the rows left unthinned for use during the winter and early spring. The ideal type for this purpose is the Greentop White; this will give a heavy crop of tender young foliage in some eight to ten weeks from the time of sowing.

Monthly Reminders

January

Work in the garden during this month is often restricted by the weather, but there are usually borders or beds which require digging or manuring when possible. Naturally, all preparatory work should be done in good time and if a greenhouse is available a start can be made here. Intending exhibitors should early in the month sow seeds of Selected Ailsa Craig and Prizetaker Leek in pans or trays, the seedlings from which should be pricked off preferably into single pots for the best results or into trays.

At the end of the month it should be safe to sow Progress or May Express Cabbage for cutting heads from late May/early June onwards. A similar operation can be carried out with Classic Cauliflower.

Both Amstel Carrot and Red Forcing Radish may be sown in a greenhouse, frames, or Dutch lights.

Asparagus beds should be well manured, but the beds should not be dug. Lay the manure on the soil for the rain to wash in.

Broad Beans may be sown in frames or under cloches, and, usually towards the end of the month, sow in the open ground, for which Aquadulce is recommended.

Lettuces: sow seeds of early-maturing varieties in pans for pricking off in trays to transplant outside as soon as the plants are large enough and have been carefully hardened off.

Peas: round-seeded hardy types may be sown in the open, selecting a warm and sunny plot; Feltham First is one of the best of this group.

Tomato: where the amateur has a heated greenhouse available with a temperature of not less than 60°F, seeds can be sown thinly in a seed-pan or box which has been filled with a good seed-compost.

February

Digging and manuring the land should be continued as weather permits; the spring frosts will help to pulverize the soil.

In southerly parts of the country, round-seeded Peas such as Feltham First should be sown, as well as Colossal Broad Bean.

Tomato seed can be sown in a greenhouse with some heat; and and early Lettuce such as Fortune, could be sown thinly in trays for subsequent pricking-out, and, in due course, planting outdoors in a warm sheltered spot.

Brussels Sprouts seed for early picking can be sown in a greenhouse or frame; and usually towards the end of the month a seed-bed can be prepared for sowing Parsnip and Onion.

Garlic bulbs can be broken up into cloves for planting about 1 in. deep. Allow 6 in. between plants and 12 in. between the rows. These distances are also suitable for Onion Sets and Shallots which can be planted at the end of the month.

Capsicum, Chili, and Egg Plant can be sown now or in March, in heat. Parsley can be sown at the end of the month.

Radishes, to be mild and tender and of good shape, must grow quickly, but seed could be grown now in a frame of an early variety such as Red Forcing.

Rhubarb which requires dividing should be taken up and replanted in rich moist soil, every piece to have only one good eye. Do not pull sticks the first year from freshly-planted crowns.

Spinach can be sown in favourable soil and weather conditions, choosing a round-seeded variety.

Turnip: a variety such as Milan could be sown in frames—not in the open ground.

March

This is frequently the busiest month of the year in the garden, as so many different crops can be sown if the soil is in a friable condition and the weather reasonably settled.

Sow Brussels Sprouts and Onions outdoors, and, towards the end of the month, early Carrots, Radish, and Lettuce.

Peas (wrinkled or marrowfat types), Broad Beans, and Spinach can be sown.

Complete plantings of Onion sets, Shallots, and Garlic.

Plant early Potatoes, preferably in a sheltered position, at the end of the month.

At the end of the month, all Cabbage sown in January/February and Cauliflower sown September/October and raised under glass should be hardened off for April transplanting.

If Cabbages are showing the effects of the winter, top dress with Sulphate of Ammonia or Nitro Chalk to bring back colour and encourage strong growth.

The exhibitors' Onion-bed should be prepared for planting out seedlings next month.

Cabbage: sow two or three varieties in a seed-bed to provide Cabbages for cutting during the late summer and throughout the autumn.

Celery and Celeriac can be sown early in the month under glass for providing the first sticks of the season.

Chives can be divided and re-planted in a fresh spot.

Garlic may still be planted.

Herbs of many kinds may be sown or divided.

Leek: sow the main crop in a seed-bed in rich, well-prapared soil. Lettuce: plant out whatever seedlings you have available, including those raised under glass; also winter hardy types which were sown in September.

Onion: the plants already raised in boxes or trays should be removed to cold frames. Keep the frames closed at first but give air with increasing freedom as the time approaches for transfer to the open ground.

Parsnip: sow the main crop in shallow drills in well-prepared, deeply-dug soil.

Spinach: sow a round-seeded type successionally as required.

Tomato: in the southern counties, especially in favourable seasons, there is no difficulty in producing a good crop in the open ground, especially if a suitable early variety such as Leader is selected. Plants which are ready should be transferred to small thumb pots. Put them in so that the first leaves touch the rim of the pot and place them in a glass frame or warm part of the greenhouse for a few days until the roots take hold. Give each plant plenty of space and avoid a forcing temperature. A shelf in the greenhouse is a good position.

April

Plant early Potatoes and follow closely with a Maincrop variety.

Make successional sowings of wrinkled-seeded Peas, such as Early Onward, Onward, and, for exhibition, Show Perfection.

Sow dwarf Beans outdoors from mid-April, and Cabbages, Savoys, and Leeks in a seed-bed for transplanting from late June onwards.

Sow *in situ* to be thinned, Beet, Perpetual Spinach, Kohlrabi, and Carrots.

Further sowings of Broad Beans may be made.

Cabbage and Cos Lettuce are by now more successful if sown in the open ground and thinned out.

Clean the ground after winter crops and prepare it for Maincrop Potatoes and successional sowings of other subjects.

Plant Asparagus crowns without delay.

Sow Climbing French Beans towards the end of the month.

Cucumber: for a summer crop to be grown in the greenhouse, the beds should be made up towards the end of the month. Strong healthy plants are required and these should be raised by sowing the seeds in pots a month in advance. Prior to planting the beds, the temperature of the house should be raised to 80°F for one whole day. Shade the plants at first, but this can be dispensed with as growth increases.

Herbs: Chervil, Fennel, and other flavouring and medicinal Herbs are best sown at this time. Rich soil is not required but the position should be dry and sunny.

Leek may be sown again if the earlier sowing is insufficient or has failed.

Onions: plants raised under glass in January or February should be ready for planting out under favourable conditions about mid-April.

Onions for pickling should be sown thickly on poor ground, well firmed. Do not thin the plants, to encourage early ripening and the production of small bulbs.

Parsley may be sown in quantity for summer and autumn supplies. Thin as soon as possible and give each plant plenty of room.

Peas may be sown in succession.

Turnip may be sown in quantity.

Marrow: an early sowing can be made in pots in readiness for planting out when the weather is suitable.

Winter Greens: a sowing of Kale can be made for transplanting and, if a supply is required in the spring, a further sowing can be made in May.

May

Further sowings in succession can be made if required of Peas, Dwarf Beans, Carrots, and Beet.

Sow Runner Beans now (in the north, delay sowing until the latter part of the month). Space the plants well and feed with potash fertilizer when they have reached 9 in. high.

After mid-May, sow Cauliflowers and Broccoli for autumn and winter cutting, and ensure that when the young plants are transplanted they do not receive a check.

Brussels Sprouts: raised under glass and March-sown outdoors; when finally transplanted—the ideal spacing is $2\frac{1}{2}$ ft. square.

Sweet Corn may be sown about mid-May, selecting a variety such as Sutton's First of All F_1 Hybrid for the earliest crop.

Celery trenches should be prepared.

Ridge Cucumber: seeds may be sown direct in the open ground about the middle of the month, but it is also possible to sow in pots in a greenhouse or frame from mid-April to plant out at the end of May.

Gourd of Pumpkin: for the production of early fruit it is advisable to sow under glass for planting on prepared beds, but for the main crop seeds can be sown on the actual bed.

Melon: seeds can be sown now for frame cultivation and the Cantaloupe type such as Ogen, for growing under cloches.

New Zealand Spinach can be sown in the open ground during the early part of the month; thin to about 3 ft apart each way.

Tomato: by about the third week in May the plants intended for the open ground should be hardened off.

Turnip may be sown in succession.

Marrow, for early cropping from the open ground, raise strong plants in pots and put them out on rich warm beds as early as the season and district will permit. Guard against late frosts and keep a careful watch for slug attacks.

June

Plant outdoor Tomatoes early in the month. If a normal variety, put the stakes in, first making the holes; add a little general fertilizer, and water the plants well before planting. Plant bush varieties similarly except that stakes are not required.

Transplant Brussels Sprouts and Ridge Cucumbers raised in pots or boxes.

Sow short rows of Lettuces every two or three weeks to maintain a continuity of cutting through to September. Sow Lettuce in the open ground and thin out, but do not transplant.

Make successional sowings of Globe Beet and Champion Scarlet Horn Carrot.

Cabbage and Savoy plants for autumn/winter cropping can be transplanted.

In the south early Potatoes are ready for lifting and the final earthing of the Maincrop should be completed.

From mid-June transplant Celery into the specially-prepared trenches, whilst Celeriac should be planted on the flat and given a good watering.

Beans, Dwarf and Runner, may be sown about the middle of the month to supply young pods when those from the early sowings are past.

Capsicums may be planted out in a sunny sheltered position.

Cucumbers for pickling may be sown on the ridges.

Melon: for the final crop in houses, sow and grow plants in pots until the house has been prepared. The growth should be pushed forward to ensure ripe fruit before the end of September.

Potato: as a precautionary measure against blight, Potato foliage should be sprayed with Bordeaux Mixture.

July

Watch staked Tomato plants and remove side shoots as they appear. As soon as the bottom truss is set, feed with a well-balanced fertilizer.

Complete transplanting Cauliflower by the middle of the month.

Make greater use of 'catch-crops' in the vegetable garden. Sow such crops *in situ* not later than the third week. Thin as soon as the seedlings are large enough and do not transplant. The following subjects are suitable: Pea, Kelvedon Wonder; Dwarf Bean, Masterpiece; Beet, Little Ball; Cabbage, Earliest; Cauliflower, Classic; Carrot, Champion Scarlet Horn; Lettuce, Avondefiance, Webb's Wonderful, or Little Gem; Marrow, Smallpak; Onion, Cocktail or White Lisbon (for salads); Radish, China Rose; and Turnip, Golden Ball.

Celery: continue to plant out in showery weather.

Garlic and Shallot may be taken up in suitable weather, and it may be necessary to complete the ripening under cover.

Leeks should be planted out in dry soil; plant them in trenches as prepared for Celery.

Parsley: sow now for winter use.

Spinach: sow Greenmarket Spinach to withstand the winter.

Tomato: for a winter supply of fruit, sow towards the end of July.

Turnips can be sown in quantity during this month.

Winter Greens of all kinds should be planted out freely in good soil.

August

Early in the month, transplant plants of Sprouting Broccoli, winter Cauliflowers, Scotch Kale, and Leek, raised from outdoor sowing.

Sow Radish and Onion White Lisbon.

Early in the month, sow Cabbage in a seed-bed for spring cutting; make final transplanting in late September/early October.

Sow Greentop White Turnip for Turnip greens.

Outdoor Tomatoes will be carrying a good crop of fruit, and to prevent disease spoiling the crop spray with Bordeaux Mixture.

Lift, thoroughly dry, and store Shallots and Onions.

Sow Solidity Onion to produce a crop next year.

Sow seed of Japanese types of Onions such as Kaizuka Extra Early and Express Yellow F_1 Hybrid mid to late August in rows 15 in. apart; do not transplant but thin in the spring to 2–3 in. apart in the rows.

September

The last sowings of Radish can be made outdoors.

Sow in the open ground Valdor Cabbage Lettuce for spring cutting.

Transplant autumn-sown Cabbages.

Earth-up Celery.

Remove outer ring of foliage from Celeriac to encourage the roots to swell.

At the end of the month, remove green trusses of outdoor Tomatoes and finish ripening them in frames or under cover.

Harvest maincrop Onion bulbs.

Lift and store maincrop Potatoes.

Cauliflower seed of a suitable variety such as Arcturus or Classic can be sown towards the end of the month, pricked off into trays or into frames where they will be wintered; keep the plants hardy by giving air at every opportunity. These plants should be planted out in the open during March–April.

Celery: continue to earth up.

Parsley: the latest sowing will require thinning.

Spinach: in favourable localities Winter Spinach can be sown in the first half of the month to make a good plant before winter. Thin the plants that are ready to about 6 in. apart.

October

Complete transplanting August-sown Cabbage.

Potatoes: complete the lifting of these.

Celery: give a further earthing-up. Put a light raffia tie round the Celery stems to prevent any soil from falling into the hearts of the plants.

Beet, Carrot, Salsify, and Turnip: towards the end of the month, mature crops of these should be lifted and stored.

Parsnips may be dug all the winter as required and, although a slight frost will not injure them when left in the ground, some protection will be necessary during very severe weather.

Rhubarb for forcing should be lifted and laid aside in a dry cool place exposed to the weather. This gives the root a check, which in some degree prepares them for forcing.

November

Broad Beans: sow a frost-hardy variety outdoors such as Aquadulce, which should crop at the end of May.

Peas: Meteor may be sown under cloches.

Carrot: A suitable variety such as Amstel can be sown in frames and successional sowings can be made every three or four weeks until March.

Sea Kale should be lifted for forcing.

December

Broad Beans, Aquadulce, and Peas, Meteor or Feltham First may be sown in the open ground.

Radish: an early variety such as Red Forcing can be sown in a heated greenhouse.

VEGETABLES AND SALADS ALL THE YEAR ROUND

The actual distances allowed between the rows or plants will depend on the variety grown and the method of cultivation.

	Sow Under Glass	Sow in the Open	Plant or Transplant	Final Distances Between Plants in Rows	Distance Between Rows	Season of Use
Asparagus	—	April	April	15–18 in.	1½ in.	May and June
Bean						
Broad	January	Nov. to April	—	6–9 in.	double rows 2–3 ft apart	June to Aug.
Dwarf French	July to April	April–July	—	8–12 in.		All year round
Runner	April	May and June	—	9–12 in.	5 ft	July to Oct.
Climbing French	July to April	End Apr.–June	—	9–12 in.	4–5 ft	All year round
Beet	Feb. and Mar.	March to July	—	4–6 in.	12–15 in.	All year round
Broccoli (Sprouting)	—	May and June	July and August	1½ ft	2 ft	Sept. to May
Brussels Sprouts	February	March and April	May and June	2½ ft	2½ ft	Oct. to March
Cabbage						
Spring sown	Late Jan. and Feb.	April to May	May to July	1½ ft	20 in.	All year round
Autumn sown	—	Early August	Sept. to March	1–1½ ft	1½–2 ft	All year round
Carrot	Nov. to Feb.	March to July	—	4–6 in.	12 in.	All year round
Cauliflower						
Summer Heading	Late Sept. or Jan.–Feb.	—	April	1½ ft	2 ft	June to July
Autumn Heading	—	Late April–May	July	1½ ft	2 ft	Aug. to Dec.
Winter Heading	—	May	July	2 ft	2 ft	Jan. to June
Celeriac	March	April	May and June	1 ft	1½ ft	Oct. onwards
Celery	March	April	May and June	6–9 in.	4 ft	Oct. to Feb.
Chicory	—	May and June	—	9 in.	1½ ft	Oct. to May
Corn Salad	—	Feb. to Oct.	—	6 in.	6 in.	June onwards
Cress	Sept. to March / Feb. to Oct.	March to Sept.	—	—	—	All year round
Cucumber						
Ridge	April and May	—	May and June	2½ ft	4 ft	All year round
Endive	March	April to July	—	12 in.	15 in.	Aug. and Sept.
Kale	Jan. and Feb.	April to May	June and July	1½–2 ft	2 ft	All year round
Leek	—	Feb. and March	April to July	6–12 in.	1½ ft	Nov. to April
Lettuce	August to March	March to Sept.	—	9–12 in.	1 ft	Oct. to March / All year round

The actual distances allowed between rows or plants will depend on the variety grown and method of cultivation.

	Sow Under Glass	Sow in the Open	Plant or Transplant	Final Distances Between Plants in Rows	Distance Between Rows	Season of Use
Marrow	Feb. to April	May	May and June	2 ft	4–5 ft	May to Oct.
Melon	Jan. onwards	—	—	—	—	Sum. and Aut.
Melon Canteloupe	April/May	May	—	3 ft	—	August/Sept.
Mustard	Sept. to March	March to Sept.	—	—	—	All year round
Onion	Dec. to Feb.	March to July August	April February	4–15 in.	1–2 ft	All year round
Onion sets	—	Jan. to March	4 in.	12–15 in.	August onwards	All year round
Parsley	February	March to July	—	6–12 in.	1 ft	Nov. to March
Parsnip	—	Feb. and April	—	12 in.	1½ ft	
Pea						
Round-seeded	Nov. to July	Nov. to March	—	2 in.	dwarfs 2 ft	April to Oct.
Wrinkled		March to July	—	2 in.	tall 4–5 ft	April to Oct.
Pumpkin (Gourd)	April	May	May	3 ft	4 ft	April to Oct.
Radish	Oct. to Feb.	March to Sept.	—	—	—	All year round
Rhubarb		—	Winter	3–4 ft	3–4 ft	Spring–Autumn
Salsify		April and May	—	9–12 in.	15 in.	Oct. to March
Savoy Cabbage		April to May	June and July	1–2 ft	1–2 ft	Sept. to March
Scorzonera		April and May	—	9–12 in.	15 in.	Oct. to March
Shallot		—	Jan. and Feb.	6–9 in.	12 in.	July to March
Spinach						
Summer	—	Feb.–May	—	9–12 in.	12–15 in.	May to Jan.
Winter	—	July to Sept.	—	9–12 in.	12–15 in.	July to April
Perpetual	—	April to July	—	9–12 in.	15 in.	late autumn/winter
Swede	—	May	—	9–12 in.	1½ ft	July to Sept.
Sweet Corn	April	April and May	May	18 in.	2½ ft	All year round
Tomato	Dec. to April	—	End of May (outdoors)	15–18 in.	1½–2 ft	All year round
Turnip (Garden)	July to August Jan. and Feb.	April to August	—	4–6 in.	12–15 in.	All year round

In judging collections of vegetables, the following scale of points will be used:

	Maximum points		*Maximum points*
Artichokes, Globe	15	Fennel, Florence	15
Artichokes, Jerusalem	10	Garlic	12
Beans, Broad	15	Herbs	5
Beans, Dwarf French	15	Kales	12
Beans, Stringless	15	Kohl rabi	12
Beans, Runner	18	Leeks	20
Beans, Climbing, other than Runner	15	Lettuces	15
Beet	15	Marrows, including Squashes	10
Broccoli, Sprouting and Coloured-headed	15	Mushrooms	12
Brussels Sprouts	15	Mustard or Rape	5
Cabbages, Green	15	Onions	20
Cabbages, Red	15	Onions, Pickling	10
Cabbages, Savoy	15	Onions, Green Salad	5
Capsicums and Chilies	15	Parsnips	20
Cardoons	15	Peas	20
Carrots	20	Potatoes	20
Cauliflowers, including White Heading 'Broccoli'	20	Pumpkins	10
Celeriac	15	Radishes	10
Celery	20	Salsify	12
Chicory, Heads (Chicons)	15	Scorzonera	12
Chives	5	Sea Kale	16
Corn Salad (Lambs' Lettuce)	5	Sea Kale Beet	12
Courgettes	10	Shallots	12
Cress	5	Spinach	12
Cress, American or Land	5	Spinach, New Zealand	12
Cucumbers	18	Spinach Beet	12
Dandelion, Blanched	5	Swedes	15
Egg Plant Fruits (Aubergines)	15	Sweet Corn	15
Endive	15	Tomatoes	18
		Turnips	15